Selected Titles in This Series

AMS/IP

Studies in
Advanced
Mathematics

Volume 12

The Bieberbach Conjecture

Sheng Gong

American Mathematical Society · International Press

Shing-Tung Yau, Managing Editor

1991 *Mathematics Subject Classification*. Primary 30C50.

This book is a revised translation of *The Bieberbach Conjecture*, Science Press, 1989, in Chinese. Permission has been granted by Science Press to reuse material from the original book translated into English and incorporated into this new volume.

Library of Congress Cataloging-in-Publication Data
Kung, Sheng, 1930–
    The Bieberbach conjecture / Sheng Gong.
       p. cm. — (AMS/IP studies in advanced mathematics ; v. 12)
    Includes bibliographical references and index.
    ISBN 0-8218-0655-6 (alk. paper)
    1. Bieberbach conjecture. I. Title. II. Series.
QA331.7.K85   1999
515′.9—dc21

99-26584
CIP

To my wife Huiyi

CONTENTS

FOREWORD

If $f(z)$ is a univalent holomorphic function on the unit disc, $D = \{z : |z| < 1\}$, in the complex plane, we may add normalization conditions, $f(0) = 0$ and $f'(0) = 1$. Thus $f(z)$ has the Taylor expansion $f(z) = z + a_2 z^2 + a_3 z^3 + \cdots + a_n z^n + \cdots$, on D. The set of all such functions forms a normal family S.

In 1916, Bieberbach conjectured: If $f \in S$, then $|a_n| \leq n$ holds true for $n = 2, 3, \cdots$ The equality holds if and only if $f(z)$ is the Koebe function $\frac{z}{(1-z)^2}$ or one of its rotations. The conjecture was not completely solved until 1984 by de Branges. That is, mathematicians spent 68 years solving this simple-looking conjecture.

During these 68 years, there were a huge number of papers discussing this conjecture and its related problems. For example, when S. D. Bernardi listed the bibliography of univalent functions, 4282 papers had been published up to 1981. No doubt, a high percentage of these papers are related to this conjecture. Moreover, during this period, many very nice books were published that systematically presented the known theory of univalent functions. Among those books are four especially nice ones by the following authors: Duren, Goluzin, Hayman and Pommerenke. These are listed in the references.

After de Branges proved this famous conjecture, I wrote and published in 1989 a small book in Chinese titled "The Bieberbach Conjecture," presenting the history of related coefficient problems and de Branges' proof. This is the English translation of my small book with many changes. In particular, it includes some results related to several complex variables. Anybody who has completed the standard material in a one year graduate complex analysis course can easily understand this small book.

Several people have been very helpful in publishing the English edition of this book. I am greatly indebted to Professor S. T. Yau for encouraging me to translate the Chinese edition of this book to English. Also I am deeply indebted to Professor Carl H. FitzGerald for writing a wonderful preface and giving me lots of important suggestions. It is a great pleasure to thank Dr. Carolyn Thomas and Dr. Weigi Gao who made many useful suggestions for mathematics and for improving the English throughout the text.

Finally, I would like to take this opportunity to express my sincere thanks to the Department of Mathematics, University of California, San Diego, for their hospitality in providing me with a stimulating environment, where I was able to complete both the Chinese edition and the English edition of this small book.

Sheng Gong   Feb. 1998

PREFACE

The dramatic story of the Bieberbach Conjecture illustrates the creation of mathematics. Made in 1916, this conjecture stood as a challenge to complex analysis for sixty-eight years. During that time, many mathematicians made contributions to mathematics of complex variables in their efforts to solve this problem. For example, M. Schiffer brought calculus of variation technique into complex analysis. C. Löwner used some of Lie's ideas to find a way to represent the functions involved as solutions to certain partial differential equations. W. Kaplan brought attention to the class of close-to-convex mappings; and M. Reade showed that the conjecture was true for this large class. And many others made impressive advances in complex analysis in their efforts to solve the problem. When the final winning assault was made on the conjecture, it was clearly manifest that a magnificent piece of mathematics had been discovered; and it was clear that earlier work had laid a foundation for that success. Thus, this history of the Bieberbach Conjecture shows some ways in which mathematicians continue to build the science of mathematics.

The initial interest in the Bieberbach Conjecture came from the completion of an earlier program. In the first decade of the twentieth century, mathematicians had studied the analytic functions $p(z) = 1 + 2c_1z + 2c_2z^2 + \cdots$ on the unit disk $\{z : |z| < 1\}$ such that the real part of $p(z)$ is positive. A very satisfactory theory was developed. In particular, the bounds $|c_n| \leq 1$ were proved for all positive integers n. These bounds are sharp since for each positive n,

$$p(z) = \frac{1+z}{1-z} = 1 + 2z + 2z^2 + 2z^3 + 2z^4 + \cdots$$

shows that the upper bound is reached. More generally, a characterization of the coefficients of positive real part functions was found.

With the successful analysis of the class of positive real part functions, it was natural to consider other classes of analytic functions. One obvious candidate was the class S of functions $f(z) = z + a_2z^2 + a_3z^3 + \cdots$ which are analytic and one to one on the unit disk. (The letter S is used for the German *Schlicht* since the Rieman surface is "simple".) The Koebe function is an interesting example of a function in S. The function is

$$K(z) = \frac{z}{(1-z)^2} = z + 2z^2 + 3z^3 + 4z^4 + \cdots.$$

It takes the unit disk onto the plane minus the negative real axis from $-\frac{1}{4}$ to minus infinity. Bieberbach showed that $|a_2| \leq 2$. In a footnote, he indicated the general expectation that $|a_n| \leq n$ for $n = 2, 3, 4, \cdots$; and furthermore, for each n, the only the functions which attain the upper bound are the Koebe function and its rotations $K_\theta(z) = e^{-i\theta}K(e^{i\theta}z)$.

The problem quickly became a focus of complex analysis. When in 1923 Löwner presented his proof that $|a_3| \leq 3$, Bieberbach shook his hand and assured him that he had joined the "realm of the immortals". Also Bieberbach suggested that Löwner put a "one" at the end of the title of the paper; the next installment would include the solution for all n. But, of course, much happened after first paper before Löwner's theory became a tool in de Branges' proof of the Bieberbach Conjecture.

The eminent mathematician, Professor Sheng Gong, tells this story of the Bieberbach Conjecture by presenting a large sample of the mathematical results it inspired. In particular, his survey includes de Branges' proof of the conjecture. To his original Chinese version of this book, Professor Gong has added a presentation of L. Weinstein's simplification of the de Branges' proof, H. Wilf's comments on Weinstein's proof and some others.

Professor Sheng Gong has had a dynamic career. As a student he studied with the internationally respected mathematician, Hua Lou-keng. Through the years, Gong's principal employer has been the important University of Science and Technology of China. (There was a hiatus during the Cultural Revolution to acquire first hand knowledge of rural agriculture.) He held many administrative positions; in particular, he became the vice president in charge of foreign affairs and personnel at USTC. Also Professor Gong has visited several American universities, including the University of California at San Diego.

The mathematical interests of Professor Gong have been in one and several complex variables. Indeed, he is one of the founders of modern complex analysis in China. Four of his Chinese books include *Harmonic Analysis on Classical Groups, The Integral of Cauchy Type on the Ball, Convex and star-like mappings in several complex variables* and *The Bieberbach Conjecture.* Each of these books has been translated into English and published for the benefit of mathematicians in the West.

Professor Gong has used his expertise as a writer, a teacher and a research mathematician to create an attractive, readable monograph. This work is accessible to those who know the standard material in a one year graduate complex analysis course. Care has been taken to present the work in as self-contained a form as possible. Each theorem presented in worthwhile in itself. And, as a collection, these results have the additional interest of being a case study in the development of mathematics.

Carl H. FitzGerald

July 1994 at UCSD

CHAPTER I

INTRODUCTION

$1.1. Some Classical Results

Let Ω be a domain in the complex plane $\mathbb{C}$, and let $f(z)$ be a single-valued holomorphic function on Ω. The function $f(z)$ is called *univalent* if for any distinct points z_1 and z_2 in Ω, $f(z)$ takes on different values, i.e., $f(z_1) \neq f(z_2)$. $f(z)$ is called *locally univalent* if for any $z_0 \in \Omega$, $f(z)$ is univalent in some neighborhood of z_0.

By the Riemann Mapping Theorem, for any simply connected domain Ω in $\mathbb{C}$ with more than one boundary point, for any point $z_0 \in \Omega$, there exists a univalent holomorphic function $f(z)$ which maps Ω onto the unit disk $D = \{z \in \mathbb{C} : |z| < 1\}$, and $f(z_0) = 0$, $f(z_0) > 0$.

In this small book, we mainly study univalent holomorphic functions on the unit disk D.

The following basic facts about the unit disk D are well known.

1. *Group of holomorphic automorphism* of D is

$$(1.1.1) \qquad \text{Aut}\,(D) = \left\{ e^{i\alpha} \frac{z - z_0}{1 - \bar{z}_0 z} \;:\; z_0 \in D,\; \alpha \in \mathbb{R} \right\}.$$

The fractional linear transformation

$$(1.1.2) \qquad \varphi(z) = e^{i\alpha} \frac{z - z_0}{1 - \bar{z}_0 z}$$

maps z_0 to 0.

2. The *Bergman kernel function of D* is

$$(1.1.3) \qquad K(z, \bar{\zeta}) = \frac{1}{\pi(1 - z\bar{\zeta})^2}$$

where $z \in D$, $\zeta \in D$.

1

3. The *Bergman metric* of D is

(1.1.4) $$ds^2 = \frac{\partial^2}{\partial z \partial \bar{z}} \log K(z, \bar{z}) = \frac{2|dz|^2}{(1 - |z|^2)^2}.$$

Usually, we call the Bergman metric of D as *Poincaré-Bergman metric*.

It is easy to verify that the Poincaré-Bergman metric is invariant under $\text{Aut}(D)$. That is, if $w = \varphi(z) \in \text{Aut}(D)$, then

$$ds_z^2 = \frac{2|dz|^2}{(1 - |z|^2)^2} = ds_w^2 = \frac{2|dw|^2}{(1 - |w|^2)^2}.$$

If $f(z)$ is a univalent holomorphic function on D, then $f'(z) \neq 0$ for all $z \in D$. We may add the normalization conditions $f(0) = 0$ and $f'(0) = 1$. Thus $f(z)$ has the Taylor expansion

(1.1.5) $$f(z) = z + a_2 z^2 + a_3 z^3 + \cdots + a_n z^n + \cdots, \quad |z| < 1.$$

The set of all such functions form a normal family S.

The *Koebe function*

(1.1.6) $$K(z) = \frac{z}{(1 - z)^2} = z + 2z^2 + 3z^3 + \cdots + nz^n + \cdots$$

plays an important role in S. It maps the unit disk onto the complex plane minus the ray on the negative real axis that extends from $\frac{-1}{4}$ to infinity. If θ is any real number, then $e^{-i\theta} K(e^{i\theta} z) \in S$, and it maps the unit disk onto the complex plane minus a ray that extends from $\frac{-1}{4} e^{-i\theta}$ to infinity. The ray points directly away from the origin. When θ is fixed, such a function is called a *rotation* of the Koebe function.

All normalized locally univalent functions in D form a family. We denote it by $\mathcal{S}$.

Let S_0 be a subfamily of $\mathcal{S}$. For any $f \in S_0$, let $g(z) = f(\varphi(z))$, where $\varphi(z) \in \text{Aut}(D)$, then $w(z) = \frac{g(z) - g(0)}{g'(0)} \in \mathcal{S}$. If $w(z) \in S_0$ again for any $\varphi(z) \in \text{Aut}(D)$, then we call S_0 as a *linear invariant family*. This idea was introduced by Pommerenke [2].

For example, S is a linear-invariant family, etc.

Then we have the following theorem. (cf. Pommerenke [2], Gong-Zheng [1])

Theorem 1.1.1. *If S_0 is a linear-invariant family, $f(z) \in S_0$ and has expansion (1.1.5), then*

(1.1.7) $$\left| \log \frac{f'(z)}{\sqrt{K(z, \bar{z})/K(0, 0)}} \right| \leq C(S_0) \log \frac{1 + |z|}{1 - |z|}$$

holds, where $C(S_0) = \sup\{|a_2| \; : \; f \in S_0\}$. *When* $z \in D, z \neq 0$, *the equality holds only for the function which* $|a_2| = C(S_0)$.

Proof. Let $\varphi_\zeta \in AutD$, and $\varphi_\zeta(0) = \zeta \in D$, and let $g(w) = f\left(\varphi_\zeta(w)\right)$. We expand $g(w)$ at $w = 0$, then

$$g(w) = g(0) + g'(0)w + \frac{1}{2}g''(0)w^2 + \cdots.$$

It is easy to verify that

$$g'(0) = f'(\zeta)\varphi'_\zeta(0) \quad \text{and} \quad g''(0) = f''(\zeta)(\varphi'_\zeta(0))^2 + f'(\zeta)\varphi''_\zeta(0).$$

Normalizing $g(w)$, we obtain a function

$$F(w) = (g(w) - g(0))(g'(0))^{-1} =$$

$$w + \frac{1}{2}\left[\frac{f''(\zeta)}{f'(\zeta)}\varphi'_\zeta(0) + \frac{\varphi''_\zeta(0)}{\varphi'_\zeta(0)}\right]w^2 + \cdots.$$

Then $F(w) \in S_0$ again. Let

$$(1.1.8) \qquad c_2 = \frac{1}{2}\left[\frac{f''(\zeta)}{f'(\zeta)}\varphi'_\zeta(0) + \frac{\varphi''_\zeta(0)}{\varphi'_\zeta(0)}\right],$$

we have

$$(1.1.9) \qquad \frac{d}{d\zeta}\log f'(\zeta) = \frac{2c_2}{\varphi'_\zeta(0)} - \frac{\varphi''_\zeta(0)}{(\varphi'_\zeta(0))}.$$

By the property of the Bergman Kernel function, we know

$$(1.1.10) \qquad K(w, \bar{w}) = |\varphi'_\zeta(w)|^2 K(\varphi_\zeta(w), \overline{\varphi_\zeta(w)}).$$

Differentiating both sides of (1.1.10) with respect to w, then let $w = 0$, we have

$$0 = \left.\frac{\partial}{\partial w}K(w, \bar{w})\right|_{w=0}$$

$$= K(\zeta, \bar{\zeta})\overline{\varphi'_\zeta(0)}\varphi''_\zeta(0) + |\varphi'_\zeta(0)|^2\left.\frac{\partial}{\partial w}K(\varphi_\zeta(w), \overline{\varphi_\zeta(w)})\right|_{w=0}$$

$$= K(\zeta, \bar{\zeta})\overline{\varphi'_\zeta(0)}\varphi''_\zeta(0) + |\varphi'_\zeta(0)|^2\frac{\partial}{\partial\zeta}K(\zeta, \bar{\zeta}) \cdot \varphi'_\zeta(0).$$

It follows

(1.1.11)
$$\frac{\varphi_\zeta''(0)}{\varphi_\zeta'(0)} = -\frac{\frac{\partial}{\partial \zeta} K(\zeta, \bar{\zeta})}{K(\zeta, \bar{\zeta})} \varphi_\zeta'(0).$$

Substituting (1.1.11) into (1.1.9), we have

(1.1.12)
$$\frac{d}{d\zeta} \log f'(\zeta) = \frac{2c_2}{\varphi_\zeta'(0)} + \frac{\frac{\partial}{\partial \zeta} K(\zeta, \bar{\zeta})}{K(\zeta, \bar{\zeta})}.$$

By the definition of Bergman kernel function, we know that

$$\frac{\partial}{\partial \bar{\zeta}} K(\zeta, \bar{\zeta}) = \overline{\frac{\partial}{\partial \zeta} K(\zeta, \bar{\zeta})}.$$

For a fix $z \in D$, let $\zeta = tz, 0 \le t \le 1$, then

$$\frac{d}{dt} K(\zeta, \bar{\zeta}) = \frac{\partial}{\partial \zeta} K(\zeta, \bar{\zeta}) \frac{d\zeta}{dt} + \frac{\partial}{\partial \bar{\zeta}} K(\zeta, \bar{\zeta}) \frac{d\bar{\zeta}}{dt} = 2Re \left\{ \frac{\partial}{\partial \zeta} K(\zeta, \bar{\zeta}) z \right\}.$$

But we know that
$$\Im m \left\{ \frac{\partial}{\partial \zeta} K(\zeta, \bar{\zeta}) z \right\} = 0.$$

Thus
$$\frac{\partial}{\partial \zeta} K(\zeta, \bar{\zeta}) = \frac{1}{2z} \frac{d}{dt} K(\zeta, \bar{\zeta}).$$

(1.1.12) becomes

$$\frac{\partial}{\partial \zeta} \log f'(\zeta) = \frac{2c_2}{\varphi_\zeta'(0)} + \frac{1}{2} \frac{d}{d\zeta} \log K(\zeta, \bar{\zeta}),$$

that is,

(1.1.13)
$$\frac{d}{d\zeta} \log \frac{f'(\zeta)}{\sqrt{K(\zeta, \bar{\zeta})}} = \frac{2c_2}{\varphi_\zeta'(0)}.$$

Integrating both sides of (1.1.13) with respect to t from 0 to 1, we obtain

$$\frac{1}{z} \log \frac{f'(z)/f'(0)}{\sqrt{K(z, \bar{z})/K(0, 0)}} = \pi \int_0^1 \frac{c_2 dt}{\varphi_\zeta'(0)}.$$

Thus

$$(1.1.14) \qquad \left| \log \frac{f'(z)}{\sqrt{K(z,\bar{z})/K(0,0)}} \right| = 2 \left| \int_0^1 \frac{c_2 z dt}{\varphi'_\zeta(0)} \right| \leq 2C(S_0) \int_0^1 \frac{|z| dt}{|\varphi'_\zeta(0)|}.$$

Let $w = 0$ in (1.1.10), then

$$K(0,0) = |(\varphi'_\zeta(0))|^2 K(\zeta, \bar{\zeta}).$$

By (1.1.3), we have $|\varphi'_\zeta(0)| = 1 - |\zeta|^2$, substituting it into (1.1.14) we have (1.1.7).

For equality to hold it is necessary to have $|c_2| = C(S_0)$ on the radius from 0 to z, where c_2 is defined by (1.1.8), and hence in particular at 0. This means that $|a_2| = C(S_0)$.

Corollary 1.1.1. *If S_0 is a linear invariant family, for any $f \in S_0$, the inequality*

$$(1.1.15) \qquad \sqrt{\frac{K(z,\bar{z})}{K(0,0)}} \left(\frac{1-|z|}{1+|z|} \right)^{C(S_0)} \leq |f'(z)| \leq \sqrt{\frac{K(z,\bar{z})}{K(0,0)}} \left(\frac{1+|z|}{1-|z|} \right)^{C(S_0)}$$

holds, where $C(S_0) = \sup\{|a_2| : f \in S\}$.

When $z \subset D, z \neq 0$, the equality holds only for the function which $|a_2| = C(S_0)$.

More explicitly, (1.1.15) is equivalent to

$$(1.1.16) \qquad \frac{(1-|z|)^{C(S_0)-1}}{(1+|z|)^{C(S_0)+1}} \leq |f'(z)| \leq \frac{(1+|z|)^{C(S_0)-1}}{(1-|z|)^{C(S_0)+1}}$$

by (1.1.3).

Corollary 1.1.2. *If S_0 is a linear-invariant family, then $C(S_0) \geq 1$.*
Proof. If $C(S_0) < 1$, then by (1.1.16), for any $f \in S_0$, and any $z \in D$, $\min |f'(z)| \to \infty$ as $z \to \partial D$. This is impossible.

Corollary 1.1.3. *If S_0 is a linear-invariant family. For any $f \in S_0$, the inequality*

$$\frac{1}{2C(S_0)} \left[1 - \left(\frac{1-|z|}{1+|z|} \right)^{C(S_0)} \right] \leq |f(z)|$$

$$(1.1.17) \qquad\qquad \leq \frac{1}{2C(S_0)} \left[\left(\frac{1+|z|}{1-|z|} \right)^{C(S_0)} - 1 \right]$$

holds, where $C(S_0) = \sup\{|a_2| : f \in S_0\}$.

When $z \in D, z \neq 0$, equality holds only for the function which $|a_2| = C(S_0)$.

Proof.   Let $f \in S_0, z = re^{i\theta}, 0 < r < 1$. Since $f(0) = 0$, we have

$$f(z) = \int_0^r f'(\rho e^{i\theta}) e^{i\theta} d\rho.$$

By (1.1.16),

$$|f(z)| \leq \int_0^r |f'(\rho e^{i\theta})| d\rho \leq \int_0^r \frac{(1+\rho)^{C(S_0)-1}}{(1-\rho)^{C(S_0)+1}} d\rho$$

$$= \frac{1}{2C(S_0)} \left[\left(\frac{1+r}{1-r} \right)^{C(S_0)} - 1 \right].$$

It is the right-hand side inequality of (1.1.21).

To prove the left-hand side inequality of (1.1.21), let $m(r)$ denote the minimum of $|f(z)|$ on $|z| = r$. f maps z-plane to w-plane. The image of $|z| < r$ contains the disk $|w| < m(r)$. Therefore, there exists a curve γ from 0 to $|z| = r$, such that

$$m(r) = \int_\gamma |f'(z)||dz|.$$

Since γ intersects all circle $|z| = \rho < r$, the lower bound for $|f'(z)|$ leads to the desired estimate

$$m(r) \geq \int_0^r \frac{(1-\rho)^{C(S_0)-1}}{(1+\rho)^{C(S_0)+1}} d\rho = \frac{1}{2C(S_0)} \left[1 - \left(\frac{1-|z|}{1+|z|} \right)^{c(S_0)} \right].$$

Corollary 1.1.4.   If S_0 is a linear-invariant family. The image of the unit disk D under a mapping $f \in S_0$ contains the disk centred at 0 with radius $\frac{1}{2C(S_0)}$.

The mapping $f \in S_0$ which contains the disk centred at 0 with radius $\frac{1}{2C(S_0)}$ and not contains a larger disk centred at 0, is the mapping for which $|a_2| = C(S_0)$.

Now we consider the class S again. It is a linear-invariant family. Closely connected to the family S is the function family Σ, which consists of univalent meromorphic function on $\Delta = \{z \in \mathbb{C} : |z| > 1\}$ of the form

$$(1.1.18) \qquad g(z) = z + b_0 + b_1 z^{-1} + b_2 z^{-2} + \cdots$$

i.e. univalent functions holomorphic outside the unit disk except for a pole with residue 1 at infinity. The family Σ is obviously a normal family also.

All functions $g \in \Sigma$ with the property that $g(z) \neq 0$ for all $z \in \Delta$ form a family Σ'. It is clear that any function in Σ can be transformed into a function in Σ' by an appropriate translation.

If $f \in S$, then

$$(1.1.19) \qquad g(z) = \left\{ f\left(\frac{1}{z}\right) \right\}^{-1} = z - a_2 + (a_2^2 - a_3)z^{-1} + \cdots$$

is in Σ'. Conversely, any function g in Σ' can be transformed into a function $f \in S$ by inverting the preceding transformation. Therefore the preceding transformation establishes a one-to-one correspondence between S and Σ'.

All functions $g \in \Sigma$ with the property that g maps $|z| > 1$ onto the complex plane minus a set with two-dimensional Lebesgue measure zero form a family $\widetilde{\Sigma}$.

The following theorem is classical, but important.

Theorem 1.1.2(Gronwall Area Principle). *For any function $g(z) \in \Sigma$ defined by* (1.1.18), *the inequality*

$$(1.1.20) \qquad \sum_{n=1}^{\infty} n|b_n|^2 \leq 1$$

holds. The equality holds if and only if $g \in \widetilde{\Sigma}$.

Proof. Let E be the complement of the image of g in $\mathbb{C}$ and C_r be the image of the set $|z| = r \, (r > 1)$ under g. Since g is univalent, C_r is a simple closed curve bounding a domain E_r. Obviously $E_r \supset E$. By the Green Theorem, the area of E_r is

$$A_r = \frac{1}{2i} \int_{c_r} \bar{w} dw = \frac{1}{2i} \int_{|z|=r} \overline{g(z)} g'(z) dz$$

$$= \frac{1}{2} \int_0^{2\pi} \{ re^{-i\theta} + \sum_{n=0}^{\infty} \bar{b}_n r^{-n} e^{in\theta} \}.$$

$$\left\{ 1 - \sum_{v=1}^{\infty} v b_v r^{-v-1} e^{-i(v+1)\theta} \right\} re^{i\theta} d\theta$$

$$= \pi \left\{ r^2 - \sum_{n=1}^{\infty} n |b_n|^2 \right\}$$

for $r > 1$. As $r \to 1+$, A_r tends to the outer measure of E,

$$m(E) = \pi \{ 1 - \sum_{n=1}^{\infty} n |b_n|^2 \}.$$

Since $m(E) \geq 0$, the theorem follows.

Corollary 1.1.5.   *If $g \in \Sigma$, then $|b_1| \leq 1$. The equality holds if and only if*

$$g(z) = z + b_0 + \frac{b_1}{z}, \quad |b_1| = 1.$$

$g(z)$ maps Δ onto a domain with the complement of a line segment of length 4.

The proof is obvious.

Theorem 1.1.3(Bieberbach).   *For any function $f \in S$ defined by (1.1.5), we have $|a_2| \leq 2$. The equality holds if and only if f is the Koebe function defined by (1.1.6) or one of its rotations.*
Proof.   Since $f \in S$ and $f(z) = 0$ only at the origin $z = 0$, a single-valued branch of the square root of $f(z^2)$ may be chosen, then

$$g(z) = \left\{ f \left(\frac{1}{z^2} \right) \right\}^{-\frac{1}{2}} = z - \frac{a_2}{2} z^{-1} + \cdots \in \Sigma.$$

By Corollary 1.1.5, $|a_2| \leq 2$, and the equality holds if and only if $g(z) = z - \frac{e^{i\theta}}{z}$. It follows that $f(z) = \frac{z}{(1-e^{i\theta}z)^2} = e^{-i\theta} K(e^{i\theta} z)$, a rotation of the Koebe function.

Theorem 1.1.3 was proved by L. Bieberbach in 1916. It is the origin of the Bieberbach conjecture.

From Theorem 1.1.3 and Corollary 1.1.1, Corollary 1.1.3, Corollary 1.1.4, we have

Theorem 1.1.4(Distortion, Growth and Covering Theorem for S). *For any function $f \in S$, the inequalities*

(1.1.21)
$$\frac{1-r}{(1+r)^3} \leq |f'(z)| \leq \frac{1+r}{(1-r)^3},$$

(1.1.22)
$$\frac{r}{(1+r)^2} \leq |f(z)| \leq \frac{r}{(1-r)^2}$$

hold, and

$$f(D) \supset D\left(0, \frac{1}{4}\right)$$

where $D(0,\frac{1}{4})$ means a disk centred at 0 with radius $\frac{1}{4}$. When $z \in D, z \neq 0$, the equality holds if and only if f is the Koebe function or one of its rotations.

A domain is *convex* if the line segment connecting any two points in the domain is contained in the domain.

If $f \in S$, with expansion (1.1.5), and f maps the unit disk D onto a convex domain, then we call f a univalent convex holomorphic function. All univalent convex holomorphic functions in S form a subfamily, which we denote by C. Obviously C is a linear-invariant family.

Let $f \in C$, then for any positive integer m, $\ y(z) = \frac{1}{m} \sum_{j=1}^{m} f\left(c^{\frac{i2j\pi}{m}} z\right)$

$\in f(D)$. It is easy to verify

$$g(z) = a_m z^m + \cdots .$$

Let $f^{-1}(g(z)) = v(z)$, then $v(0) = 0$, and $|v(z)| < 1$ when $|z| < 1$. We have $g(z) = f(v(z))$. By the definition of f and g, the equality

$$a_m z^m + \cdots = v(z) + a_2 v(z)^2 + \cdots$$

holds. Comparing the lowest order of z on both sides, it is

$$a_m z^m = \frac{1}{2\pi} \int_0^{2\pi} v(ze^{i\theta}) e^{-im\theta} d\theta.$$

We have $|a_m z^m| < 1$. Let $|z| \to 1$, it follows $|a_m| \leq 1$.

Combining Corollary 1.1.1, Corollary 1.1.3 and Corollary 1.1.4, we have the following theorem. (Löwner [2])

Theorem 1.1.5(Distortion, Growth and Covering Theorem for C)   *For any function $f \in S$ with expansion (1.1.5), the inequalities*

(1.1.23) $$|a_n| \leq 1 \quad n = 1, 2, \cdots$$

(1.1.24) $$\frac{1}{(1+r)^2} \leq |f'(z)| \leq \frac{1}{(1-r)^2},$$

(1.1.25) $$\frac{r}{1+r} \leq |f(z)| \leq \frac{r}{1-r}.$$

hold, and

$$f(D) \supset D(0, \frac{1}{2})$$

where $D(0, \frac{1}{2})$ means a disk centred at 0 with radius $\frac{1}{2}$. When $z \in D, z \neq 0$, the equality holds if and only if $f = \frac{z}{1-z}$ or one of its rotations.

$1.2. The Bieberbach Conjecture

Bieberbach formulated the following conjecture after he proved $|a_2| \leq 2$ in 1916 (L. Bieberbach [1]).

Bieberbach Conjecture. *For any function $f \in S$ of the form (1.1.5), the inequality $|a_n| \leq n$ holds for all $n = 2, 3, \cdots$. The equality holds if and only if $f(z)$ is the Koebe function or one of its rotations.*

K. Löwner introduced the parametric representation method in 1923 and proved that $|a_3| \leq 3$ (Löwner [1]). This method is the cornerstone of the proof the Bieberbach conjecture by L. de Branges, and it is one of the main methods in geometric function theory. A series of important result can be proved using this method. We will discuss this method in some detail in the next chapter.

Little progress was made for thirty-two years after the proof of $|a_3| \leq 3$ until Garabedian and Schiffer proved $|a_4| \leq 4$ in 1955, using a variational method (P. R. Garabedian and M. Schiffer [1]). Their proof was long and complicated. However, Z. Charzynski and M. Schiffer [1] found another proof of $|a_4| \leq 4$ in 1960 by using the Grunsky Inequality. Their method is surprisingly simple, and drew people's attention to the Grunsky Inequality. We will state this important inequality here and give more detail of it in Chapter 3.

Let $g \in \Sigma$. Then

$$(1.2.1) \qquad \log \frac{g(z) - g(\zeta)}{z - \zeta} = -\sum_{n=1}^{\infty} \sum_{k=1}^{\infty} \gamma_{nk} z^{-k} \zeta^{-n}, \quad |z| > 1, \ |\zeta| > 1$$

is holomorphic for all $|z| > 1$ and $|\zeta| > 1$. Moreover, we have the following inequality.

Grunsky inequality. *For any positive integer N and any N complex numbers $\lambda_1, \cdots, \lambda_N$, the inequality*

$$(1.2.2) \qquad \left| \sum_{n=}^{N} \sum_{k=1}^{N} \gamma_{nk} \lambda_n \lambda_k \right| \leq \sum_{n=1}^{N} \frac{1}{n} |\lambda_n|^2$$

holds.

(1.2.2) is called the *weak Grunsky Inequality*. It is equivalent to the *strong Grunsky Inequality*

$$(1.2.3) \qquad \sum_{k=1}^{\infty} k \left| \sum_{n=1}^{N} \gamma_{nk} \lambda_n \right|^2 \leq \sum_{n=1}^{N} \frac{1}{n} |\lambda_n|^2.$$

They are also equivalent to the *generalized weak Grunsky Inequality*

$$(1.2.4) \qquad \left| \sum_{n=1}^{N} \sum_{k=1}^{N} \gamma_{nk} \lambda_n \mu_k \right|^2 \leq \sum_{n=1}^{N} \frac{1}{n} |\lambda_n|^2 \sum_{k=1}^{N} \frac{1}{k} |\mu_k|^2.$$

In 1968, Pederson and Ozawa proved $|a_6| \leq 6$ independently, using the Grunsky inequality. We omit the detail of the proof (cf. R. N. Perderson [1], M. Ozawa [1]. S. Gong [2]).

The Grunsky Inequality has many applications. For example, from (1.2.4), we immediately have

$$\left| \sum_{n=1}^{\infty} \sum_{k=1}^{\infty} \gamma_{nk} \sum_{i=1}^{N} \lambda_i z_i^{-k} \sum_{j=1}^{N} \mu_j \zeta_j^{-n} \right|^2$$

$$\leq \left\{ \sum_{k=1}^{\infty} \frac{1}{k} \left| \sum_{i=1}^{N} \lambda_i z_i^{-k} \right|^2 \right\} \left\{ \sum_{k=1}^{\infty} \frac{1}{n} \left| \sum_{j=1}^{N} \mu_j \zeta_j^{-n} \right|^2 \right\}$$

where $\lambda_1, \cdots, \lambda_N$, $\mu_1, \cdots, \mu_N$ are $2N$ arbitrary complex numbers, and $z_1, \cdots, z_N$, $\zeta_1, \cdots, \zeta_N$ are $2N$ points outside the unit disk. It is equivalent to

$$\left| \sum_{i=1}^{N} \sum_{j=1}^{N} \lambda_i \mu_j \log \frac{g(z_i) - g(\zeta_j)}{z_i - \zeta_j} \right|^2 \leq$$

(1.2.5)

$$\left\{ -\sum_{i=1}^{N} \sum_{j=1}^{N} \lambda_i \lambda_j \log \left(1 - \frac{1}{z_i \bar{z}_j} \right) \right\} \cdot \left\{ -\sum_{i=1}^{N} \sum_{j=1}^{N} \mu_i \mu_j \log \left(1 - \frac{1}{\zeta_i \bar{\zeta}_j} \right) \right\}.$$

The proof of $|a_5| \leq 5$ is more difficult. It requires to use the generalized form of the Grunsky Inequality, the Garabedian-Schiffer Inequality. We omit the detail of the proof. (cf. Pederson-Schiffer [1]). Until de Branges proved the Bieberbach Conjecture in full in 1984, we could only prove $|a_n| \leq n$ for $n \leq 6$.

People did know, however, long before de Branges' proof, that the Bieberbach Conjecture was true for some special families of functions contained in S.

We already mentioned in §1.1, that for any $f \in C$ with expansion (1.1.5), $|a_n| \leq 1$. Moreover, we will give the criterion of $f \in C$.

If f maps the unit disk D onto a convex domain, then f maps any domain $D_r = \{z \in \mathbb{C} : |z| < r\}$ ($r < 1$) onto a convex domain. Indeed, for any two points z_1 and z_2, $|z_1| \leq |z_2| \leq r$, let $w_1 = f(z_1), w_2 = f(z_2)$ and $w_0 = tw_1 + (1-t)w_2, 0 < t < 1$. Since f maps $|z| < 1$ onto a convex domain, there exists a point z_0 in $|z| < 1$ such that $f(z_0) = w_0$. We need to show that $|z_0| < r$.

The function $g(z) = tf(z_1 z/z_2) + (1-t)f(z)$ is holomorphic on unit disk D, and $g(0) = 0, g(z_2) = w_0$. The function $h(z) = f^{-1}(g(z))$ is well-defined in D, and $h(0) = 0, |h(z)| \leq 1$. By Schwarz lemma, $|h(z)| \leq |z|$. That is, $|z_0| = |h(z_2)| \leq |z_2| < r$.

f maps $|z| = r < 1$ onto a convex curve. As z traces out the circle $|z| = r$ in the positive direction, the slope of the tangent line to the image curve is non-decreasing. This can be written analytically as

$$\frac{\partial}{\partial \theta} \left(\arg \left\{ \frac{\partial}{\partial \theta} f(re^{i\theta}) \right\} \right) \geq 0,$$

or,

$$\Im m \left\{ \frac{\partial}{\partial \theta} \log[ire^{i\theta} f'(re^{i\theta})] \right\} \geq 0,$$

that is,

$$\Re e \left\{ 1 + \frac{zf''(z)}{f'(z)} \right\} \geq 0, \ |z| = r.$$

Conversely, it's true also.

Lemma 1.2.1. *Let $f(z) \in S$, then $f(z) \in C$ if and only if*

(1.2.6)
$$\Re e \left\{ 1 + \frac{zf''(z)}{f'(z)} \right\} \geq 0$$

holds for any $z \in D$.

A domain is *starlike with respect to origin* if any ray emanating from the origin to any other point of the domain lies entirely in the domain.

If $f \in S$, with expansion (1.1.5), and f maps the unit disk D onto a starlike domain (with respect to origin), then we call f a univalent starlike holomorphic function. All univalent starlike holomorphic functions in S form a subfamily, we denote it by S^*. S^* is not a linear-invariant family. We will give the criterion of $f \in S^*$.

Let Ω be the image of D under the univalant starlike holomorphic function $f(z)$. Then for any fixed $t, 0 < t < 1, tf(z)$ belongs to Ω. There exists a function $w(z)$, holomorphic on D with $|w(z)| \leq 1$ such that $tf(z) = f[w(z)]$. Clearly, $w(0) = 0$. By Schwarz lemma $|w(z)| \leq |z|$. Choose $0 < \rho < 1$, and consider the function $g(z) = f(\rho z)$, we have

$$tg(z) = tf(\rho z) = f[w(\rho z)] = f\left[\rho \frac{w(\rho z)}{\rho} \right].$$

Obviously, $w_1(z) = \frac{w(\rho z)}{\rho}$ satisfies $|w_1(z)| \leq |z|$. Thus $tg(z) = g[w_1(z)]$, $|w_1(z)| \leq |z|, 0 < t < 1$. It follows $g(z)$ is a starlike mapping. That is $f(z)$ maps $|z| < \rho$ onto a starlike domain. $f(z)$ maps $|z| = \rho$ into a starlike curve, the boundary of a starlike domain. When $z = \rho e^{i\theta}$ traces out the circle $|z| = \rho, f(z) = Re^{i\phi}$ must also move in only one direction. Otherwise a ray from the origin will cut it at more than one point. It follows that

$$\frac{\partial}{\partial \theta} \arg\{f(z)\} = \frac{\partial \phi}{\partial \theta} \geq 0.$$

Conversely, if the above condition is satisfied, then the curve is a starlike curve. However, $\log f(z) = \log R + i\phi$, the above condition is equivalent to

$$\Im m \left\{ \frac{\partial \log f(z)}{\partial \theta} \right\} \geq 0.$$

Fix ρ, then

$$\frac{\partial}{\partial \theta} = \rho e^{i\theta} \frac{d}{dz} = iz \frac{d}{dz}.$$

Hence $\Im m \left\{ iz \frac{d}{dz} \log f(z) \right\} = \Re e \left\{ \frac{zf'(z)}{f(z)} \right\} \geq 0.$

Lemma 1.2.2. *Let* $f(z) \in S$, *then* $f(z) \in S^*$ *if and only if*

$$(1.2.7) \qquad\qquad \Re e \left\{ \frac{zf'(z)}{f(z)} \right\} \geq 0$$

holds for any $z \in D$.

Lemma 1.2.3 (Alexander). *Let* $g(z) \in S$, *then* $g(z) \in S^*$ *if and only if* $g(z) = zh'(z)$, *where* $h(z) \in C$.
Proof. It is a consequence of Lemma 1.2.1 and Lemma 1.2.2.
   From Lemma 1.2.3 and Theorem 1.1.5, we have the following result (cf. Nevanlinna [1]).

Theorem 1.2.1(Nevanlinna). *If* $f \in S^*$ *with the expansion* (1.1.5), *then* $|a_k| \leq k$ *holds for all* $k = 2, 3, \cdots$. *The equality holds if and only if* f *is the Koebe function or one of its rotations.*
   That means the Bieberbach conjecture holds true when $f \in S^*$.
   No doubt, the family of univalent convex holomorphic functions C, and the family of univalent starlike holomorphic functions S^*, are two most important subfamilies of the family S.
   In 1952, W. Kaplan generalized the concept of starlike function to that of a close-to-convex function. A holomorphic function $f(z)$ in D is *close-to-convex* if there exists a starlike function $g(z)$, such that for any $z \in D$, the inequality

$$(1.2.8) \qquad\qquad \Re e \left\{ z \frac{f'(z)}{g(z)} \right\} > 0$$

holds. This is clearly a generalization of (1.2.7).
   By Lemma 1.2.3, if $h(z)$ is a convex function, then $g(z) = zh'(z)$ is a starlike function. The condition $\Re e \left\{ z \frac{f'(z)}{g(z)} \right\} > 0$ is equivalent to $\Re e \left\{ \frac{f'(z)}{h'(z)} \right\} > 0$ for some convex function $h(z)$.

We will show that a close-to-convex function is a univalent function. Let convex function $h(z)$ map D onto a domain H. Let $\phi(w) = f(h^{-1}(w))$, then for $w \in H$,

$$\Re e \phi'(w) = \Re e \left\{ \frac{f'(z)}{h'(z)} \right\} > 0.$$

For any two points $w_1, w_2 \in H$,

$$\Re e \frac{\phi(w_2) - \phi(w_1)}{w_2 - w_1} = \int_0^1 \Re e\{\phi'(w_1 + t(w_2 - w_1))\}dt > 0.$$

Hence $\phi(w)$ is univalent in H, and $f(z) = \phi(h(z))$ is univalent in $|z| < 1$. We have the following result (Reade [1]).

Corollary 1.2.1.   *If $f(z)$ is close-to-convex with expansion (1.1.5), then $|a_n| \leq n$ holds true for all $n = 2, 3, \cdots$.*

To prove Corollary 1.2.1, we need the following Lemma.

Lemma 1.2.4(Carathéodory).   *If $f(z) = 1 + c_1 z + c_2 z^2 + \cdots + c_n z^n + \cdots$ is holomorphic on the unit disk D, and has positive real part, then*

$$|c_n| \leq 2, \quad n = 1, 2, \cdots$$

This inequality is sharp for each n.
Proof. $\Re e f \geq 0$ if and only if $|1 + f(z)| \geq |1 - f(z)|$. Let $g(z) = \frac{f(z)-1}{f(z)+1} = \frac{c_1}{2} z + \cdots$, then $|g(z)| \leq 1$, and $g(0) = 0$. By Schwarz lemma $|g(z)| = \left| \frac{c_1}{2} z + \cdots \right| \leq |z|$. It implies $|c_1| \leq 2$.
Let $\eta_k, \; k = 1, \cdots, n$ be the distinct n-th roots of unit, then

$$\Re e \left(\frac{1}{n} \sum_{k=1}^{n} f(\eta_k z^{\frac{1}{n}}) \right) = \Re e\{1 + c_n z + \cdots\} \geq 0.$$

Using the previous result, we have $|c_n| \leq 2, \; n = 2, 3, \cdots$.
The function $\varphi(z) = \frac{1+z}{1-z} = 1 + 2 \sum_{n-1}^{\infty} z^n$ shows this inequality is sharp for each n.
Now we are going to prove Corollary 1.2.1.
Let $z \frac{f'(z)}{g(z)} = c_0 + c_1 z + \cdots$, and $g(z) = b_1 z + \cdots$, then

$$n a_n = c_0 b_n + \sum_{v=1}^{n-1} c_{n-v} b_v, \quad n = 2, 3, \cdots.$$

By (1.2.8) and Lemma 1.2.4, we have $|c_v| \leq 2|c_0|$ for $v \geq 1$. By Theorem 1.2.1, $|b_v| \leq v|b_1|$. Using the relation $c_0 b_1 = 1$, we obtain

$$n|a_n| \leq n + 2 \sum_{v=1}^{n-1} v = n^2.$$

It follows that $|a_n| \leq n$.

Of course there are other subfamilies of S for which the Bieberbach Conjecture is true. Here we only show one more. (cf. Rogosinski [1], Dieudonné [1], Szasz [1])

Theorem 1.2.2. *If $f \in S$ and if all the coefficients a_n in (1.1.5) are real numbers, then $|a_n| \leq n$ holds for all $n = 2, 3, \cdots$.*
Proof. Let $z_1 = re^{i\varphi}, z_2 = re^{-i\varphi}, \varphi \neq 0, 0 < r < 1$, then

$$\frac{f(z_1) - f(z_2)}{z_1 - z_2} = 1 + \sum_{n=2}^{\infty} a_n \frac{z_1^n - z_2^n}{z_1 - z_2} = 1 + \sum_{n=2}^{\infty} a_n r^{n-1} \frac{\sin n\varphi}{\sin \varphi} \neq 0$$

Since all a_n are real, the expression

$$\Phi(r, \varphi) = 1 + \sum_{n=2}^{\infty} a_n r^{n-1} \frac{\sin n\varphi}{\sin \varphi}$$

is real and non-zero when $\varphi \neq 0$. We know that $\Phi(0, \varphi) = 1$, hence $\Phi(r, \varphi) \geq 0$ when $0 \leq r < 1$, and $\varphi \neq 0$. It is easy to verify that

$$2 \sin^2 \varphi \Phi(r, \varphi) = 1 + a_2 r \cos \varphi$$
$$+ (a_3 r^2 - 1) \cos 2\varphi + (a_4 r^2 - a_2) r \cos 3\varphi$$
$$+ \cdots + (c_n r^2 - c_{n-2}) r^{n-3} \cos(n-1)\varphi + \cdots \geq 0.$$

Fix $r, 0 < r < 1$, consider the function

$$F(z) = 1 + c_2 rz + (a_3 r^2 - 1)z^2 + (a_4 r^2 - a_2)rz^3 + \cdots$$
$$+ (c_n r^2 - c_{n-2})r^{n-3}z^{n-1} + \cdots$$

then $\Re e F(z) \geq 0$. By Lemma 1.2.4, we have

$$|a_2| \leq 2, \ |a_3 r^2 - 1| \leq 2, \ |a_4 r^2 - a_2|r \leq 2, \cdots, |a_n r^2 - a_{n-2}|r^{n-1} \leq 2, \cdots$$

Let $r \to 1$, and by induction, we have proved the Theorem.

The family of convex functions is contained in the family of starlike functions, and the family of starlike functions is contained in the family of close-to-convex functions. A detailed discussion of many of these special families of functions can be found in the book by A. W. Goodman [1].

We have the following general estimate of the coefficients in (1.1.5). The first important result was obtained by J. E. Littlewood in 1925.

Theorem 1.2.3   *For any $f \in S$ with expansion* (1.1.5), *the following inequality holds*

$$|a_n| < en, \qquad n = 2, 3, \cdots .$$

Littlewood's proof is based on an estimate of the mean of modulus of f
(1.2.9)

$$M_p(r, f) = \left\{ \frac{1}{2\pi} \int_0^{2\pi} |f(re^{i\theta})|^p \, d\theta \right\}^{\frac{1}{p}}, \qquad 0 < p < \infty, \quad 0 < r < 1.$$

Lemma 1.2.5.   *For any $f \in S$,*

$$(1.2.10) \qquad\qquad M_1(r, f) \leq \frac{r}{1 - r}, \qquad 0 < r < 1.$$

The Littlewood Theorem is a corollary of the above lemma.
Since

$$a_n = \frac{1}{2\pi i} \int_{|z|=r} \frac{f(z)}{z^{n+1}} \, dz = \frac{1}{2\pi} \int_0^{2\pi} e^{-in\theta} r^{-n} f(re^{i\theta}) \, d\theta,$$

we have $|a_n| r^n \leq \frac{1}{2\pi} \int_0^{2\pi} |f(re^{i\theta})| \, d\theta$, i.e., $|a_n| \leq r^{-n} M_1(r, f)$. By (1.2.10), $|a_n| \leq r^{-n+1}(1 - r)^{-1}$. The right hand side of the above inequality has its minimum when $r = 1 - \frac{1}{n}$. Thus

$$|a_n| \leq n \left(1 + \frac{1}{n-1} \right)^{n-1} < en.$$

This proves the theorem.
Now we are going to prove Lemma 1.2.5.
Let $f \in S$. Because $\frac{f(z)}{z}$ is holomorphic and $\frac{f(z)}{z} \neq 0$ when $z \in D$, we can define $g(z) = \left[\frac{f(z)}{z} \right]^{\frac{1}{2}}$, $g(0) = 1$, and subsequently, $h(z) = z(g(z^2))$. The function $h(z)$ is univalent, for $h(z_1) = h(z_2)$ implies $f(z_1^2) = f(z_2^2)$, hence $z_1 = z_2$ or $z_1 = -z_2$, the latter is ruled out because h is odd and $h \neq 0$ for $z \neq 0$. Let

$$(1.2.11) \qquad\qquad h(z) = \sqrt{f(z^2)} = \sum_{n=1}^\infty c_{2n-1} z^{2n-1},$$

then $h(z) \in S$. From the definition of $h(z)$ and from Theorem 1.1.4, we have $|h(z)| \leq \dfrac{r}{1-r^2}$, $|z| = r < 1$. The function h maps $|z| < r$ into a domain D_r that is contained in the disk $|w| < r(1-r^2)^{-1}$. Since $h(z) \in S$, the area A_r of D_r is no greater than $\pi r^2 (1-r^2)^{-2}$, the area of the disk centered at the origin with radius $r(1-r^2)^{-1}$. On the other hand

$$A_r = \int_0^{2\pi} \int_0^r |h'(\rho e^{i\theta})|^2 \rho \, d\rho \, d\theta = \pi \sum_{n=1}^{\infty} n|c_n|^2 r^{2n}.$$

The inequality $\sum_{n=1}^{\infty} n|c_n|^2 r^{2n-1} \leq \dfrac{r}{(1-r^2)^2}$ holds for $0 \leq r < 1$. Integrating this inequality from 0 to r, we have $\sum_{n=1}^{\infty} |c_n|^2 r^{2n} \leq \dfrac{r^2}{1-r^2}$. This is

$$\frac{1}{2\pi} \int_0^{2\pi} |h(re^{i\theta})|^2 \, d\theta \leq \frac{r^2}{1-r^2}.$$

Substituting r for r^2 in the above inequality, we have (1.2.10).

Estimates of the general $M_p(r,f)$, $(0 < p < \infty, 0 < r < 1)$ were obtained by H. Prawitz in 1927. He proved the following theorem.

Theorem 1.2.4 (Prawitz Theorem). *If $f(z) \in S$, then for arbitrary p, $0 < p < \infty$, we have*

$$(1.2.12) \qquad M_p^p(r,f) \leq p \int_0^r \frac{1}{t} M_\infty^p(t,f) \, dt,$$

where $0 < r < 1$, $M_\infty(r,f) = \max_{|z|=r}\{f(z)\}$. A smooth Jordan curve is a Jordan curve whose parametric representation is twice continuously differentiable.

Lemma 1.2.6 *Let C_1 and C_2 be two smooth Jordan curves containing the origin. If C_1 is in the interior of C_2, then*

$$\int_{C_1} r^p \, d\theta \leq \int_{C_2} r^p \, d\theta, \qquad 0 < p < \infty,$$

where (r, θ) are the polar coordinates of the point z.

Proof. Let D be the annulus between C_1 and C_2 and let C be its boundary. Since $d\theta = \dfrac{\partial \theta}{\partial x} \, dx + \dfrac{\partial \theta}{\partial y} \, dy$, by Green Theorem,

$$\int_{C_2} r^p \, d\theta - \int_{C_1} r^p \, d\theta = \int_C r^p \, d\theta$$

$$= p \iint_D r^{p-1} \left\{ \frac{\partial r}{\partial x} \frac{\partial \theta}{\partial y} - \frac{\partial r}{\partial y} \frac{\partial \theta}{\partial x} \right\} dx \, dy.$$

But $\dfrac{1}{r}\dfrac{\partial r}{\partial x} = \dfrac{\partial \theta}{\partial y}$, $\dfrac{1}{r}\dfrac{\partial r}{\partial y} = -\dfrac{\partial \theta}{\partial x}$, hence we have

$$\int_C r^p \, d\theta = p \iint_D r^p \left\{ \left(\frac{\partial \theta}{\partial x} \right)^2 + \left(\frac{\partial \theta}{\partial y} \right)^2 \right\} dx \, dy \geq 0.$$

This proves the lemma.

We now prove Prawitz Theorem.

Denote $w = f(re^{i\theta}) = Re^{i\phi}$. One of the Cauchy-Riemann equations of $\log f$ is

$$\frac{1}{R}\frac{\partial R}{\partial r} = \frac{1}{r}\frac{\partial \phi}{\partial \theta}.$$

If C_r is the image of $|z| = r$ under f, then

$$2\pi \frac{d}{dr} M_p^p(r,f) = \int_0^{2\pi} \frac{\partial}{\partial r} R^p \, d\theta = \frac{p}{r} \int_0^{2\pi} R^p \frac{\partial \phi}{\partial \theta} \, d\theta$$

$$= \frac{p}{r} \int_{C_r} R^p \, d\phi.$$

For any $\varepsilon > 0$, the circle $\Gamma_r : |w| = R = M_\infty(r,f) + \varepsilon$ contains the curve C_r. By Lemma 1.2.6, we have

$$\int_{C_r} R^p \, d\phi \leq \int_{\Gamma_r} R^p \, d\phi = 2\pi [M_\infty(r,f) + \varepsilon]^p.$$

Letting $\varepsilon \to 0+$, we get

$$\frac{d}{dr} M_p^p(r,f) \leq \frac{p}{r} M_\infty^p(r,f).$$

Integration of both sides of the above inequality yields Prawitz Theorem.

Lemma 1.2.5 results if we let $p = 1$ in Prawitz Theorem.

For a few decades, estimation of the mean of the modulus of functions was the main method of obtaining estimates of the coefficients. E. Landau showed that $|a_n| < \left(\frac{1}{2} + \frac{1}{\pi} \right) en$ in 1929 (E. Landau [1]). G. M. Goluzin showed that $|a_n| < \frac{3}{4} en$ in 1946 (G. M. Goluzin [6]). I. E. Bazilevich proved that

$$|a_n| < \frac{9}{4} \left(\frac{1}{\pi} \int_0^\pi \frac{\sin x}{x} \, dx + 0.2649 \right) n$$

in 1947 (I. E. Bazilevich [4]). In 1951, I. E. Bazilevich and I. M. Milin independently proved the results $|a_n| < \frac{1}{2} en + 1.51$ and $|a_n| < \frac{1}{2} en + 1.8$ (I. E. Bazilevich [5], A. Lebedev and I. M. Milin [2]).

The sharp estimate of the mean of the modulus of holomorphic univalent functions was obtained by A. Baernstein in 1974 (A. Baernstein [1]). His technique is also important in other fields. He proved that, for any $f \in S$,

$$(1.2.13) \qquad M_p(r, f) \leq M_p(r, K), \qquad 0 < p < \infty,$$

holds,where K is the Koebe function defined by (1.1.6). From this result follows the estimate $|a_n| < \frac{e}{2}n$. Baernstein's result is far more than this. In fact, he proved that if $\Phi(x)$ is a convex non-decreasing function on $-\infty < x < \infty$, then for any $f \in S$, $0 < r < 1$, we have

$$\int_0^{2\pi} \Phi(\log |f(re^{i\theta})|) \, d\theta \leq \int_0^{2\pi} \Phi(\log |K(re^{i\theta})|) \, d\theta,$$

where K is the Koebe function. Moreover if Φ is strictly convex, then the equality holds for some r if and only if f is a rotation of K.

Letting $\Phi(x) = e^{px}$ in the preceding inequality, we obtain (1.2.13).

Baernstein's proof of the theorem relies on the Baernstein star function. We omit the definition of this function and the proof of the theorem in this book. Interested readers are referred to A. Baernstein [1] or P. Duren [1], Chapter 7.

The estimate of $M_p(r, f')$, i.e., the estimate of the mean of modulus of the derivative of a function $f \in S$, is still open. It is known that for $p < \frac{1}{3}$, $M_p(r, f') \leq M_p(r, K')$ does not hold. It is conjectured to hold for $p > \frac{1}{3}$.

The first improvement of the estimate constant $\frac{e}{2}$ using a method other than the estimate of the mean of modulus was due to I. M. Milin. In 1965, he pioneered a new method, in proving that for any $f \in S$

$$|a_n| < 1.243n, \qquad n = 2, 3, \cdots.$$

See I. M. Milin [1,2]. A more detailed proof is presented in Chapter 3.

Milin's record was kept until 1972, when it was improved by C. H. FitzGerald (C. H. FitzGerald [1]). FitzGerald exponentiated the Goluzin Inequality (cf. §2.4) to obtain the following estimate:

$$|a_n| < \sqrt{\frac{7}{6}}n < 1.081n, \qquad n = 2, 3, \cdots.$$

Later, Horowitz (D. Horowitz [1,2]) followed up a suggestion made in FitzGerald(C. H. FitzGerald[1] p.367, Remark 3.2) to use the FitzGerald inequalities again and again, and obtained the more accurate estimates

$$|a_n| < \left(\frac{209}{140}\right)^{\frac{1}{6}} n < 1.0691n, \qquad n = 2, 3, \cdots,$$

and

$$|a_n| < 1.0657n, \qquad n = 2, 3, \cdots .$$

FitzGerald pointed out that his method cannot lead to an eventual proof of the Bieberbach Conjecture. Neverthless, these were the best estimations on the Bieberbach conjecture before de Branges' celebrated proof. FitzGerald's work will be presented in Chapter 2. The following table summarizes the aforementioned results:

Name	Year	$|a_n| < Cn$	
Littlewood	1923	$|a_n| < en$	$\approx 2.7183n,$
Landau	1929	$|a_n| < \left(\dfrac{1}{2} + \dfrac{1}{\pi}\right) en$	$\approx 2.2244n,$
Goluzin	1946	$|a_n| < \dfrac{3}{4}en$	$\approx 2.0388n,$
Bazilevich	1947	$|a_n| < \dfrac{9}{4}\left(\dfrac{1}{\pi}\displaystyle\int_0^\pi \dfrac{\sin x}{x}\,dx + 0.2649\right)n$	$\approx 1.9240n,$
Milin	1949	$|a_n| < \dfrac{1}{2}en + 1.80$	$\approx 1.3592n + 1.80,$
Bazilevich	1949	$|a_n| < \dfrac{1}{2}en + 1.51$	$\approx 1.3592n + 1.51,$
Milin	1964	$|a_n| < \dfrac{(e^{1.6} - 1)^{\frac{1}{2}}}{1.6}n$	$\approx 1.2427n,$
FitzGerald	1971	$|a_n| < \sqrt{\dfrac{7}{6}}n$	$\approx 1.0802n,$
Horowitz	1975	$|a_n| < \left(\dfrac{209}{140}\right)^{\frac{1}{6}}n$	$\approx 1.0691n,$
Horowitz	1977	$|a_n| < \left(\dfrac{1,659,164,137}{681,080,400}\right)^{\frac{1}{14}}n$	$\approx 1.0657n.$

This is the history of general estimates of $|a_n|$ before de Branges solved the Bieberbach Conjecture in 1984.

§1.3. The Robertson Conjecture and the Milin Conjecture

A problem that is closely connected to the Bieberbach Conjecture is the estimate of coefficients of univalent odd functions. If $f \in S$, then the function $h(z)$ defined by (1.2.11) is an odd function. J. E. Littlewood and

R. E. A. C. Paley proved in 1932 that $|c_n| < 14$, and conjectured that $|c_n| \leq 1$ (J. E. Littlewood and R. E. A. C. Paley [1]). Since

(1.3.1) $$a_n = c_1 c_{2n-1} + c_3 c_{2n-3} + \cdots + c_{2n-1} c_1,$$

the Bieberbach Conjecture would be true if the Littlewood-Paley Conjecture were true by Schwarz inequality. However, in 1933, M. Feteke and G. Szegö (M. Feteke and G. Szegö [1]) disproved the Little-Paley conjecture by proving the following sharp result (proved in §2.3):

$$|c_5| \leq \frac{1}{2} + e^{-\frac{2}{3}} = 1.013 \cdots .$$

Moreover, A. C. Schaeffer and D. C. Spencer (A. C. Schaeffer and D. C. Spencer [2]) proved in 1943 that for any $n \geq 5$, there exists a univalent odd function with real coefficients such that $|c_n| > 1$. In 1976, G. B. Leeman (G. B. Leeman [1]) proved that for univalent odd functions with real coefficients, the sharp bound for the modulus of c_7 is $\frac{1090}{1083}$, dashing the hope of a nice rational bound of the modulus of c_n. After the Littlewood-Paley estimate, K. K. Chen proved in 1933 that $|c_n| < e^2$ (K. K. Chen [1]). V. I. Levin [1] proved in 1935 that $|c_n| < 2^{\frac{1}{4}} \cdot e^{\frac{1}{2}} \cdot 3^{\frac{1}{2}} = 3.39 \cdots$. In 1951, S. Gong [1],II proved that $|c_n| < 2^{-\frac{1}{6}} \cdot e^{\frac{1}{2}} \cdot 3^{\frac{1}{2}} = 2.54 \cdots$. In 1967, I. M. Milin improved the result to $|c_n| < 1.17$. Milin's result was proved using the profound Milin Method. We will give the proof in this section (I. M. Milin [3]). In 1980, V. I . Milin [1] improved the result again to $|c_n| < 1.14$. Up to date, as we know, the best estimate of the upper bound of $|c_n|$ is $|c_n| < 1.1305$, which was given by Ke Hu [1]. We will state the proof at §3.4.

Although the Littlewood-Paley Conjecture is not true, M. S. Robertson made the following conjecture in 1936.

Robertson Conjecture. *For any odd function* $h(z) = z + c_3 z^3 + c_5 z^5 + \cdots$ *in S, the inequalities*

(1.3.2) $$1 + |c_3|^2 + \cdots + |c_{2n-1}|^2 \leq n, \qquad n = 2, 3, \cdots .$$

are true.

From (1.3.1) it is apparent that the Robertson Conjecture implies the Bieberbach Conjecture.

When $n = 2$, the Robertson Conjecture is the same as $|a_2| \leq 2$. In 1936, Robertson proved that the conjecture is true for $n = 3$, by using Löwner Method. S. Friedland proved that the conjecture is true for $n = 4$ in 1970, by using the Grunsky Inequality. The conjecture remained open for $n \geq 5$

until 1984, when de Branges proved the Milin Conjecture (to be discussed in this section), which implies the Robertson Conjecture, and hence the Bieberbach Conjecture.

In 1955, W. K. Hayman (W. K. Hayman [2]) proved the following important theorem.

Theorem 1.3.1 (Hayman Regularity Theorem). *For any $f \in S$,*

$$(1.3.3) \qquad \lim_{n \to \infty} \frac{|a_n|}{n} = \alpha \leq 1.$$

The equality holds if and only if f is the Koebe function or one of its rotations.

This theorem is based on the following lemmas.

Lemma 1.3.1   *If $f \in S$, and $M_\infty(r, f) = \max_{|z|=r} |f(z)|$, then*

$$\lim_{r \to 1} (1 - r)^2 M_\infty(r, f) = \alpha \leq 1.$$

If f is not the Koebe function or one of its rotations, then $r^{-1}(1-r)^2 M_\infty(r, f)$ is strictly decreasing on the interval $(0, 1)$ and $\alpha < 1$. Proof.   Fix $z \in D$, then for any $f \in S$

$$F(\zeta) = \frac{f\left(\frac{\zeta+z}{1+\bar{z}\zeta}\right) - f(z)}{(1 - |z|^2)f'(z)} \in S$$

since $\frac{\zeta+z}{1+\bar{\zeta}} \subset Aut(D)$. By Theorem 1.1.4 (1.1.22), we have

$$\frac{|\zeta|}{(1 + |\zeta|)^2} \leq |F(\zeta)| \leq \frac{|\zeta|}{(1 - |\zeta|)^2}.$$

Let $\zeta = -z$, then

$$\frac{|z|}{(1 + |z|)^2} \leq \left| \frac{f(z)}{(1 - |z|^2)f'(z)} \right| \leq \frac{|z|}{(1 - |z|)^2}.$$

Hence

$$\frac{1 - r}{1 + r} \leq \left| \frac{zf'(z)}{f(z)} \right| \leq \frac{1 + r}{1 - r}, \qquad |z| = r < 1.$$

The equality holds if and only if $f(z)$ is the Koebe function or one of its rotations. Therefore

$$\frac{\partial}{\partial r} \log |f(re^{i\theta})| \leq \left| \frac{f'(re^{i\theta})}{f(re^{i\theta})} \right| \leq \frac{1 + r}{r(1 - r)}.$$

Integrating the previous inequality from r_1 to r_2 $(0 < r_1 < r_2 < 1)$, we have

$$\log \left| \frac{f(r_2 e^{i\theta})}{f(r_1 e^{i\theta})} \right| \le \int_{r_1}^{r_2} \frac{1+r}{r(1-r)} \, dr = \log \frac{r_2(1-r_1)^2}{r_1(1-r_2)^2}.$$

Thus for any θ, $0 < r_1 < r_2 < 1$,

$$\frac{(1-r_2)^2}{r_2} |f(r_2 e^{i\theta})| < \frac{(1-r_1)^2}{r_1} |f(r_1 e^{i\theta})|.$$

Choose θ so that $|f(r_2 e^{i\theta})| = M_\infty(r_2, f)$, then

$$\frac{(1-r_2)^2}{r_2} M_\infty(r_2, f) < \frac{(1-r_1)^2}{r_1} |f(r_1 e^{i\theta})|$$

$$\le \frac{(1-r_1)^2}{r_1} M_\infty(r_1, f).$$

This shows that $r^{-1}(1-r)^2 M_\infty(r, f)$ is a decreasing function of r. It approaches the limit $\alpha \ge 0$ as r tends to 1. By Theorem 1.1.4, (1.1.22), $r^{-1}(1-r)^2 M_\infty(r, f) \le 1$. Hence $\alpha \le 1$. If f is the Koebe function or one of its rotations, then equality holds in the above and $\alpha = 1$. Otherwise the decreasing is strict and $\alpha < 1$.

The limit α is called the *Hayman index* of the function $f \in S$.

Lemma 1.3.2   *If the Hayman index of $f \in S$ is positive, $\alpha > 0$, then there exists a unique direction $e^{i\theta_0}$ such that*

$$\lim_{r \to 1} (1-r)^2 |f(re^{i\theta_0})| = \alpha.$$

Proof.   Let $\{r_n\}$ be an increasing sequence with limit 1. Choose θ_n, $0 \le \theta_n < 2\pi$ such that $|f(r_n e^{i\theta_n})| = M_\infty(r_n, f)$, $n = 1, 2, \cdots$. By Lemma 1.3.1, for $r < r_n$, we have

$$\alpha \le \frac{(1-r_n)^2}{r_n} |f(r_n e^{i\theta_n})| \le \frac{(1-r)^2}{r} |f(re^{i\theta})|.$$

Let θ_0 be an accumulation point of θ_n. Then

$$\alpha \le \frac{(1-r)^2}{r} |f(re^{i\theta_0})| \le \frac{(1-r)^2}{r} M_\infty(r, f).$$

When $r \to 1$, $r^{-1}(1-r)^2 M_\infty(r, f) \to \alpha$. Hence

$$\lim_{r \to 1} (1-r)^2 |f(re^{i\theta_0})| = \alpha.$$

Next we show the uniqueness of θ_0. Choosing $N = 1$, $\lambda_1 = \mu_1 = 1$, $\zeta_1 = \zeta_1$, $z_1 = \zeta_2$, $|\zeta_1| = |\zeta_2| = \rho > 1$ in (1.2.5) we have

$$1 - \rho^{-2} \leq \left| \frac{g(\zeta_1) - g(\zeta_2)}{\zeta_1 - \zeta_2} \right| \leq \frac{1}{1 - \rho^{-2}}.$$

For $f(z) = \left(g\left(\frac{1}{z} \right) \right)^{-1}$, $\rho^{-1} = r$, $\frac{1}{\zeta_1} = z_1 = re^{i\theta}$, and $\frac{1}{\zeta_2} = z_2 = re^{i\theta_0}$, the above inequalities become

$$\frac{1 - r^2}{r} |e^{i\theta} - e^{i\theta_0}| \leq \left| \frac{1}{f(re^{i\theta})} - \frac{1}{f(re^{i\theta_0})} \right|$$

$$\leq \frac{1}{|f(re^{i\theta})|} + \frac{1}{|f(re^{i\theta_0})|}.$$

If $e^{i\theta} \neq e^{i\theta_0}$ and $\alpha > 0$, the fact that $\lim_{r \to 1} (1 - r)^2 |f(re^{i\theta})| \neq 0$ leads to a contradiction.

The direction $e^{i\theta_0}$ is called the *Hayman direction*.

Using these results, Hayman proved (1.3.3). The constant α in (1.3.3) is the same α that appeared in Lemmas 1.3.1 and 1.3.2. There are many proofs for the Hayman Regularity theorem. For example, Hayman's proof (cf. W. K. Hayman [1],[2]), Milin's proof (J. M. Milin [5]). An argument of Pommerenke (Ch. Pommerenke [1]) based on FitzGerald's inequality is shorter but only yields $\overline{\lim}_{n \to \infty} \frac{|a_n|}{n} < 1$ unless $f(z)$ is Koebe function. We will give the proof of this theorem at §3.3.

By the Hayman Regularity Theorem, for any function $f \in S$, there exists a large enough constant $N(f)$ such that if $n > N(f)$ then $|a_n| \leq n$. The constant $N(f)$ depends on f, so we cannot conclude that the Bieberbach Conjecture is true for large enough n. However we may formulate the following conjecture.

Asymptotic Bieberbach Conjecture *If $f \in S$ has expansion* (1.1.5), *and if $A_n = \max_{f \in S} |a_n|$, then*

$$\lim_{n \to \infty} \frac{A_n}{n} = 1.$$

By Theorem 1.1.4, the image of D of any $f \in S$ contains a disk centered at the origin with radius $\frac{1}{4}$. Littlewood made the following conjecture.

Littlewood Conjecture   *If $f \in S$ has expansion* (1.1.5) *and $f \neq w$,*
then

$$|a_n| \leq 4|w|n, \qquad n = 2, 3, \cdots.$$

Since $|w| \geq \frac{1}{4}$, the Bieberbach Conjecture implies the Littlewood Conjecture. Furthermore, Z. Nehari (Z. Nehari [1]) proved that the Asymptotic Bieberbach Conjecture implies the Littlewood Conjecture. In fact, he proved the following result.

Let $f \in S$ with expansion (1.1.5). If w is not a value of $f(z)$ for any $|z| < 1$, then

$$|a_n| \leq 4|w|\lambda n, \qquad n = 2, 3, \cdots,$$

where $\lambda = \lim_{n \to \infty} \dfrac{A_n}{n}$, $A_n = \max_{f \in S} |a_n|$.

We omit the detail of the proof of the above result.

On the other hand, D. H. Hamilton (D. H. Hamilton [1]) proved that the Littlewood Conjecture implies the Asymptotic Bieberbach Conjecture. Therefore these two conjectures are equivalent.

The Milin Method is based on the following idea: the information one gets from the Grunsky Inequality concerns the logarithms of coefficients of a univalent function. To obtain information about the function itself, it is necessary to exponentiate the Grunsky Inequality.

The following three inequalities, due to Lebedev and Milin, are very important. They give relations between coefficients of a function and coefficients of its exponentiation. The univalency assumption for functions in these inequalities is not needed.

Let $\phi(z) = \sum_{k=1}^{\infty} \alpha_k z^k$ be an arbitrary power series with a positive radius of convergence, $\phi(0) = 0$. Define

$$(1.3.4) \qquad e^{\phi(z)} = \sum_{k=0}^{\infty} \beta_k z^k, \qquad \beta_0 = 1.$$

Then the following inequalities hold.

Lebedev-Milin Inequalities

$$(1.3.5) \qquad \sum_{k=0}^{\infty} |\beta_k|^2 \leq \exp\left\{ \sum_{k=1}^{\infty} k|\alpha_k|^2 \right\},$$

$$(1.3.6) \qquad \frac{1}{n+1} \sum_{k=0}^{\infty} |\beta_k|^2 \leq \exp\left\{ \frac{1}{n+1} \sum_{m=1}^{n} \sum_{k=1}^{m} \left(k|\alpha_k|^2 - \frac{1}{k} \right) \right\},$$

$$(1.3.7) \qquad |\beta_k|^2 \leq \exp\left\{\sum_{k=1}^{n}\left(k|\alpha_k|^2 - \frac{1}{k}\right)\right\}.$$

The proof of these inequalities will be given in Chapter 3 (N. A. Lebedev and I. M. Milin [2], I. M. Milin [3]).

Let $f \in S$, and

$$(1.3.8) \qquad \log\frac{f(z)}{z} = 2\sum_{n=1}^{\infty}\gamma_n z^n, \qquad |z| < 1.$$

If $f(z)$ is the Koebe function, then obviously $\gamma_n = \frac{1}{n}$. In 1967, Bazilevich (I. E. Bazilevich [1, 2]) proved the following inequality.

Bazilevich Inequality *Let $f \in S$, γ_n be defined by (1.3.8), and let $e^{i\theta_0}$ be the Hayman direction of f. Then*

$$(1.3.9) \qquad \sum_{n=1}^{\infty} n\left|\gamma_n - \frac{1}{n}e^{-in\theta_0}\right|^2 \leq \frac{1}{2}\log\frac{1}{\alpha},$$

where $\alpha > 0$ is the Hayman index of the function f.

This inequality shows that the closer α is to 1, the "closer" f is to the Koebe function. But the inequality $|\gamma_n| \leq \frac{1}{n}$ does not hold in general. It is true only for some special families of functions, such as the family of starlike functions.

However, Milin proved the following lemma by using the Grunsky Inequality.

Milin Lemma *For any $f \in S$, the following inequality:*

$$(1.3.10) \qquad \sum_{k=1}^{n} k|\gamma_k|^2 \leq \sum_{k=1}^{n}\frac{1}{k} + \delta,$$

holds where $\delta < 0.312$.

The proof of the Milin Lemma will be given in Chapter 3. This lemma implies the following theorem.

Theorem 1.3.2 *Let $h(z)$ be defined by (1.2.11). Then*

$$|c_n| < e^{\frac{\delta}{2}} < 1.17, \qquad n = 2, 3, \cdots.$$

Proof.   Since $h(z) = \sqrt{f(z^2)}$, $f \in S$, we have

$$\log \frac{h(\sqrt{z})}{\sqrt{z}} = \frac{1}{2} \log \frac{f(z)}{z} = \sum_{n=1}^{\infty} \gamma_n z^n,$$

i.e.,

(1.3.11) $$\sum_{n=0}^{\infty} c_{2n+1} z^n = \exp\left\{ \sum_{n=1}^{\infty} \gamma_n z^n \right\}, \qquad c_1 = 1.$$

By the third Lebedev-Milin Inequality (1.3.7) we have

(1.3.12) $$|c_{2n+1}|^2 \leq \exp\left\{ \sum_{k=1}^{n} k|\gamma_k|^2 - \sum_{k=1}^{n} \frac{1}{k} \right\}.$$

It then follows, from the Milin Lemma, that

$$|c_{2n+1}| \leq e^{\frac{\delta}{2}} < e^{0.156} < 1.17, \qquad n = 1, 2, \cdots.$$

Since the Littlewood-Paley Conjecture is not true, the value of δ cannot be zero. The following Milin Conjecture claims that $\delta = 0$ in some average sense.

Milin Conjecture.   *For any $f \in S$, let γ_n be defined by (1.3.8). Then*

(1.3.13) $$\sum_{m=1}^{n} \sum_{k=1}^{m} \left(k|\gamma_k|^2 - \frac{1}{k} \right) \leq 0, \qquad n = 1, 2, \cdots.$$

From (1.3.11) and the second Lebedev-Milin Inequality, we see that

$$\sum_{k=0}^{n} |c_{2k+1}|^2 \leq (n+1) \exp\left\{ \frac{1}{n+1} \sum_{m=1}^{n} \sum_{k=1}^{m} \left(k|\gamma_k|^2 - \frac{1}{k} \right) \right\}.$$

Hence if the Milin Conjecture is true, then

$$\sum_{k=0}^{n} |c_{2k+1}|^2 \leq n+1,$$

that is, the Robertson Conjecture is true. We have proved that the Robertson Conjecture implies the Bieberbach Conjecture.

In 1984, de Branges proved the Milin Conjecture, hence also the Robertson and Bieberbach Conjectures.

To conclude this section, we mention several related conjectures.

Let $g(z) = b_1 z + b_2 z^2 + \cdots$ be a holomorphic function in $|z| < 1$, and let $f \in S$. If the image of g is contained in the image of f, then g is said to be *subordinate* to f, denoted $g \prec f$. In other words, if $g \prec f$, then there exists a Schwarz function $w(z)$ such that $g(z) = f(w(z))$.

Littlewood (J. E. Littlewood [1]) proved the following result: if $g \prec f$, then for $0 < r < 1$, $0 < p < \infty$,

$$(1.3.14) \qquad M_p(r, g) \leq M_p(r, f),$$

where $M_p(r, g)$ is defined by (1.2.9). From this result, Rogosinski proved (W. Rogosinski [2, 3]):

$$\sum_{n=1}^{N} |b_n|^2 \leq \sum_{n=1}^{N} |a_n|^2, \qquad N = 1, 2, \cdots.$$

and the inequality

$$\sum_{n=1}^{N} |b_n|^p \leq \sum_{n=1}^{N} |a_n|^p$$

does not hold when $p \neq 2$. Also it is noticed that $g \prec f$ does not imply $|b_n| \leq |a_n|$. A simple counterexample is provided by $z^2 \prec z$. Rogosinski made the following conjecture.

Rogosinski Conjecture. *If $g \prec f$, and $f \in S$, then $|b_n| \leq n$, $n = 1, 2, \cdots$.*

Since $f \prec f$ is trivially true, the Rogosinski Conjecture implies the Bieberbach Conjecture. The Rogosinski Conjecture for $n = 1$ can be proved by Schwarz lemma. For $n = 2$, a proof was given by Littlewood (J. E. Littlewood [1]). Rogosinski proved that if f is a starlike function, or if f has real coefficients, then the conjecture is true (W. Rogosinski [3]). Robertson proved the conjecture for close-to-convex functions (M. S. Robertson [1]).

But the Robertson Conjecture implies the Rogosinski Conjecture, as shown below.

Let $h(z)$ be defined by (1.2.11). Since $g \prec f$, we have $g(z) = f(w(z))$. Let

$$\phi(z) = \frac{h(\sqrt{z})}{\sqrt{z}} = 1 + c_3 z + c_5 z^2 + \cdots,$$

then $(\phi(z))^2 = f(z)/z$. Thus

$$g(z) = w(z)\{1 + c_3 w(z) + c_5(w(z))^2 + \cdots\}^2.$$

Let the partial sum of the first n terms of ϕ be denoted by

$$S_n(z) = \sum_{k=1}^{n} c_{2k-1} z^{k-1}.$$

Since $w(0) = 0$, we see that

$$b_n = \frac{1}{2\pi i} \int_{|z|=r} \frac{w(z)[S_n(w(z))]^2}{z^{n+1}} \, dz.$$

Since $S_n(w(z)) \prec S_n(z)$, by the Littlewood Theorem (1.3.14), we have

$$|b_n| \le r^{-n}[M_2(r, S_n(w(z)))]^2 \le r^{-n}[M_2(r, S_n(z))]^2$$
$$= r^{-n} \sum_{k=1}^{n} |c_{2k-1}|^2 r^{2k-2}.$$

Letting $r \to 1$, we get

$$|b_n| \le \sum_{k=1}^{n} |c_{2k-1}|^2.$$

This shows that the Robertson Conjecture implies the Rogosinski Conjecture.

The Sheil-Small Conjecture lies between the Robertson and Rogosinski Conjectures.

Let $f(z) = \sum a_n z^n$, $g(z) = \sum b_n z^n$ be two power series. Then we call $h(z) = \sum a_n b_n z^n$ the *convolution* (or *Hadamard product*) of f and g, denoted $h = f * g$.

Sheil-Small Conjecture   *For any $f \in S$ and any polynomial P of degree n, the following inequality holds.*

$$\|P * f\|_\infty \le n\|P\|_\infty,$$

where $\| \; \|_\infty$ denotes the maximum modulus in $|z| \le 1$.

If we choose $P(z) = z^n$, then the Sheil-Small Conjecture becomes the Bieberbach Conjecture. It can be shown that the Sheil-Small Conjecture is between the Robertson Conjecture and the Rogosinski Conjecture. We will omit the proof here (see Sheil-Small [1]).

These seven conjectures have the following logical relationship:

Milin Conjecture $\Longrightarrow$ Robertson Conjecture $\Longrightarrow$ Sheil-Small Conjecture $\Longrightarrow$ Rogosinski Conjecture $\Longrightarrow$ Bieberbach Conjecture $\Longrightarrow$ Asymptotic Bieberbach Conjecture $\Longleftrightarrow$ Littlewood Conjecture.

De Branges proved the Milin conjecture. Hence all of the above conjectures were proved. Before de Branges' proof, all seven conjectures were open problems.

CHAPTER II

LÖWNER THEORY

§2.1. Carathéodory Kernel Convergence Theorem

Löwner theory is the first profound theory in geometric function theory. It can be used to solve many problems. It is the cornerstone in de Branges' proof of the Bieberbach Conjecture. In this section we establish the results in the theory of functions of one complex variable which are needed in this chapter, in particular, the Carathéodory Kernel Convergence Theorem. Based upon this theorem, in the next section we establish the Löwner differential equation. In §2.3, we prove $|a_3| \leq 3$ and related results by the Löwner Method. In 1972, FitzGerald established the FitzGerald Inequality by using the Löwner Method, and proved $|a_n| < \sqrt{\frac{7}{6}} \, n$, $n = 2, 3, \cdots$. These are the contents of §2.4.

Theorem 2.1.1 (Hurwitz Theorem). *Let $\{f_n(z)\}$ be a sequence of holomorphic functions in D. Assume that $\{f_n(z)\}$ locally uniformly converges to $f(z)$. Then either $f(z)$ is identically equal to zero in D, or each zero of f is the limit of a sequence of zeros of f_n. Here D is a simply connected domain bounded by a Jordan curve.*

Theorem 2.1.2. *Let $\{f_n\}$ be a sequence of univalent holomorphic functions in D. Assume that $\{f_n\}$ locally uniformly converges to $f(z)$ as $n \to \infty$. Then $f(z)$ is either univalent in D, or is a constant in D. Here D is a simply-connected domain bounded by a Jordan curve.*

Proof. Assume the theorem does not hold. Then there exist two points z_1 and z_2 such that $f(z_1) = f(z_2) = \alpha$. By Theorem 2.1.1, for large enough

33

N, there exist disjoint neighborhoods of z_1 and z_2 such that if $n \geq N$, then $f_n(z) - \alpha$ has zeros in both neighborhoods. This contradicts the assumption that f_n is univalent for each n. Thus $f(z) \equiv \alpha$.

Theorem 2.1.3 (Montel Theorem and Its Converse). *A family of holomorphic functions in a domain is a normal family if and only if it is locally bounded in the domain.*

From Theorems 1.1.4, 2.1.2 and 2.1.3, we get the following theorem.

Theorem 2.1.4. *The family S is a compact normal family.*

Proofs of Theorems 2.1.1 and 2.1.3 can be found in standard textbooks of function theory.

Let $\mathcal{P}$ denote the family of holomorphic functions $f(z)$ in $|z| < 1$ that satisfy the conditions $f(0) = 1$ and $\Re\{f(z)\} > 0$. From the discussion in §1.2, its expansion at the origin has coefficients with modulus less than or equal to 2. For this family of functions the following theorem holds.

Theorem 2.1.5 (Herglotz Representation Theorem). *If $f(z) \in \mathcal{P}$, then f can be represented by*

$$(2.1.1) \qquad f(z) = \int_0^{2\pi} \frac{e^{it} + z}{e^{it} - z}\, d\mu(t), \qquad |z| < 1,$$

where $d\mu$ is a positive probability measure.

Proof. If $f \in \mathcal{P}$, then $\mathrm{Re} f(z) = u(re^{i\theta})$ is a positive harmonic function in D, where $z = re^{i\theta}$. For $r < 1$, define

$$\mu_r(t) = \frac{1}{2\pi} \int_0^t u(re^{i\theta})\, d\theta.$$

Then μ_r is an increasing function, $\mu_r(0) = 0$, and $\mu_r(2\pi) = u(0) = 1$. By the Helly Selection Theorem, there exists an increasing subsequence r_n, $r_n \to 1$, such that $\mu_r(t)$ converges to a non-decreasing function $\mu(t)$ in $[0, 2\pi]$. By the Poisson Integral Formula, we have

$$u(r_n z) = \frac{1}{2\pi} \int_0^{2\pi} P(r, \theta - t) u(r_n e^{it}) \, dt = \int_0^{2\pi} P(r, \theta - t) \, d\mu_{r_n}(t),$$

where $P(r, \theta - t)$ is the Poisson kernel

$$\frac{1 - r^2}{1 - 2r \cos(\theta - t) + r^2} = \mathrm{Re}\left\{ \frac{1 + z e^{-it}}{1 - z e^{-it}} \right\}.$$

Letting $n \to \infty$, and using the Helly Selection Theorem, we get

$$u(re^{i\theta}) = \int_0^{2\pi} P(r, \theta - t) \, d\mu(t).$$

Since $f(0) = 1$, we have (2.1.1).

To show the uniqueness of the representation (2.1.1), we assume

$$\int_0^{2\pi} \frac{e^{it} + z}{e^{it} - z} \, d\mu(t) \equiv 0, \qquad |z| < 1.$$

Since

$$\frac{e^{it} + z}{e^{it} - z} = 1 + 2 \sum_{n-1}^{\infty} e^{-int} z^n,$$

we see that

$$\int_0^{2\pi} e^{int} \, d\mu(t) = 0, \qquad n = 0, \pm 1, \pm 2, \cdots,$$

i.e., the integral of any trigonometric polynomial with respect to $d\mu$ is zero. But any continuous periodic function can be approximated by trigonometric polynomials, and any characteristic function of a domain can be approximated in the L^1 sense by continuous periodic functions. Hence the measure is zero on every interval, and $d\mu$ is a zero measure. This proves the uniqueness of the representation.

From Theorem 2.1.5, many properties of the family $\mathcal{P}$ may be derived. For example, from (2.1.1) we get immediately that the coefficients of the expansion of $f(z)$ at the origin have modulus less than or equal to 2. Also, if $f \in \mathcal{P}$ then

(2.1.2)
$$\frac{1 - |z|}{1 + |z|} \leq |f(z)| \leq \frac{1 + |z|}{1 - |z|}.$$

Next we prove the Carathéodory Kernel Convergence Theorem.

Let $\{F_n\}$ be a sequence of domains in $\mathbb{C}$, with a common point, say, 0. Let F contain the origin and all points $w \in \mathbb{C}$ with the property that there exists a domain H such that $0 \in H$, $w \in H$ and for sufficiently large n, $H \subset F_n$. Then F is called the *kernel* of $\{F_n\}$. Thus the kernel of $\{F_n\}$ is either $\{0\}$ or a domain in $\mathbb{C}$. If all subsequences of $\{F_n\}$ have the same kernel F, then $\{F_n\}$ is *kernel convergent* to F.

Theorem 2.1.6 (Carathéodory Kernel Convergence Theorem). *Let $\{f_n(z)\}$ be a sequence of univalent holomorphic functions in $|z| < 1$ that satisfies $f_n(0) = 0$ and $f'_n(0) > 0$, and let $F_n = f_n(D)$. Then $\{f_n(z)\}$ is locally uniformly convergent in D if and only if $\{F_n\}$ is kernel convergent to F and $F \neq \mathbb{C}$. In the case of convergence, the limit function of $\{f_n\}$ maps $|z| < 1$ onto F.*

Proof. If $f_n(z)$ locally uniformly converges to $f(z)$ as $n \to \infty$, then by Theorem 2.1.2, $f(z)$ either is univalent or constant. Therefore $F \neq \mathbb{C}$. If $f(z)$ is a constant, it equals zero, since $f_n(0) = 0$. In this case, it is obvious that $\{F_n\}$ is kernel convergent to zero. If $f(z) \not\equiv 0$, then there exists a $w_0 \neq 0$, $w_0 \in f(|z| < 1)$. We may choose r such that $w_0 = f(z_0)$, $|z_0| < r < 1$. The domain $H = \{w = f(z) : |z| < r\}$ contains both 0 and w_0. Now we show that $H \subset F_n$ for sufficiently large n. Assume the contrary, then there exists a sequence $\{n_k\}$, $n_k \to \infty$ and $w_k \in H$, such that $w_k \notin F_{n_k}$. Passing to a subsequence that converges to w^* as $k \to \infty$, $w^* \in \bar{H}$, Because $f_{n_k}(z) - w_k = (f_{n_k}(z) - w_k + w^*) - w^* \neq 0$ for all $z \in D$, Theorem 2.1.1 implies $f(z) - w^* \neq 0$ for all $|z| < 1$. This contradicts $w^* \in \bar{H} \subset f(|z| < 1)$. Thus $H \subset F_n$ for sufficiently large n. Hence for $w_0 \in f(|z| < 1)$ there exists H such that $0 \in H$, $w_0 \in H$, and $H \subset F_n$ for sufficiently large n.

Conversely, assume that for any point $w_0 \neq 0$, there exists an H such that $0 \in H$, $w_0 \in H$, and for sufficiently large n, $H \subset F_n$. Then we can show that $w_0 \in f(|z| < 1)$. Indeed, $0 = f(0)$ is in the image of f. $\phi_n(w) = f_n^{-1}(w)$ is holomorphic in H and $|\phi_n(w)| < 1$. By Theorem 2.1.3, we can find a subsequence $\{\phi_n\}$ such that it is locally uniformly convergent in H. Its limit function $\phi(z)$ satisfies $\phi(0) = 0$ and $|\phi(w)| \leq 1$. Therefore for $w \in H$, $|\phi(w)| < 1$. It follows that $f_{n_\nu}(z)$ is locally uniformly convergent near $\phi(w_0)$. Since $\phi_{n_\nu}(w_0) \to \phi(w_0)$ and $w_0 = f_{n_\nu}(\phi_{n_\nu}(w_0))$, we have $w_0 = f(z_0) \in f(|z| < 1)$. Hence the kernel F of $\{F_n\}$ is $f(|z| < 1)$.

Since every subsequence $\{f_{n_\nu}\}$ of $\{f_n\}$ converges to f, $\{F_{n_\nu}\}$ has a common kernel F. Hence $\{F_n\}$ kernel converges to F.

Now we show the converse. If F_n kernel converges to F, and $F \neq \mathbb{C}$, then $\{f_n\}$ is locally uniformly convergent in $|z| < 1$.

First we show that $\{f_n\}$ is a normal family. By Theorem 1.1.4,

$$\left\{ w : |w| < \frac{1}{4} f_n'(0) \right\} \subset F_n$$

for each n. If $f_n'(0)$ is unbounded, then there is a subsequence $\{F_{n_\nu}\}$ whose kernel is $\mathbb{C}$. But we already assumed that the kernel $F \neq \mathbb{C}$, so $f_n'(0)$ is bounded. By Theorem 1.1.4 (1.1.22), $|f_n(z)| \leq f_n'(0) \dfrac{|z|}{(1 - |z|)^2}$, $(|z| < 1)$. Hence $|f_n(z)|$ is locally uniformly bounded. By Theorem 2.1.3, $\{f_n\}$ is a normal family. Next we show that $\{f_n\}$ is uniformly convergent in $|z| < 1$. Assume not, then there exist two subsequences $\{f_{m_\nu}\}$ and $\{f_{n_\nu}\}$ locally uniformly convergent to different limit functions f and g. From the first half of this proof, the kernels of $\{F_{m_\nu}\}$ and $\{F_{n_\nu}\}$ are $f(|z| < 1)$ and $g(|z| < 1)$ respectively. Since $\{F_n\}$ kernel converges to F, we have $f(|z| < 1) = g(|z| < 1) = F$. By our assumption, $f(0) = g(0) = 0$ and $f'(0) \geq 0$ and $g'(0) \geq 0$, hence, by the uniqueness part of Riemann Mapping Theorem, $f(z) = g(z)$, a contradiction. For $F = \{0\}$ the proof is obvious. Thus $\{f_n\}$ is locally uniformly convergent in $|z| < 1$. Its limit function maps $|z| < 1$ onto F.

§2.2. Löwner Differential Equation

If a function maps the unit disk $D = \{z : |z| < 1\}$ onto the complex plane minus several Jordan arcs, then it is called a *slit mapping*. If only a single Jordan arc is missed, then the mapping is called a *single slit mapping*. Of course the slit extends to infinity. The Löwner Method relies on the denseness of single slit mappings in S.

Theorem 2.2.1. *For any function $f \in S$, there exists a sequence of single slit mappings $f_n \in S$, such that f_n converges uniformly to f on every compact subset of D, i.e., f_n locally uniformly converges to f.*

Proof. We need to show that for any function $f \in S$, and any $\varepsilon > 0$ and $0 < \rho < 1$, there exists a single slit mapping $g \in S$ such that $|f(z) - g(z)| < \varepsilon$ for $|z| \leq \rho < 1$. The theorem is then proved by choosing $\{\varepsilon_n\}$ and $\{\rho_n\}$ such that $\varepsilon_n \to 0$, $\rho_n \to 1$. Since f can be approximated uniformly on compact sets by $f(rz)/r$ $(0 < r < 1)$, which maps D onto a domain bounded by analytic Jordan curves, we may assume that f maps D onto a domain bounded by analytic Jordan curves.

Let f map D onto R, a domain bounded by an analytic Jordan curve C. Let Γ_n be a Jordan curve from infinity to a point w_0 on C followed by a part of C between w_0 and another point w_n. Let D_n be the complement of Γ_n, and f_n the map from D to D_n with $f_n(0) = 0$, $f_n'(0) > 0$. Choose the end point w_n so that $\Gamma_n \subset \Gamma_{n+1}$, $w_n \to w_0$. Then R is the kernel of the sequence $\{D_n\}$. By the Carathéodory Kernel Convergence Theorem (Theorem 2.1.6), f_n locally uniformly converges to f in D. It follows that $f_n'(0) \to f'(0) = 1$. Hence $f_n/f_n'(0) \in S$ is a sequence of single slit mappings, which locally uniformly converges to f.

Now we establish the Löwner differential equation. We first give a standard parametrization of Γ.

Assume $f \in S$ maps D onto the complex plane minus a Jordan curve Γ from a finite point w_0 to infinity. Denote this domain by R. Assume $w = \psi(t)$, $0 \leq t < T$ is a continuous parametrization of Γ, with $\psi(0) = w_0$, $\psi(s) \neq \psi(t)$ when $s \neq t$. Let Γ_t be the part of Γ that run from $\psi(t)$ to infinity, and R_t the complement of Γ_t. Then $R_0 = R$, and $R_s \subset R_t$ when $s < t$. Define

$$g(z,t) = \beta(t)\{z + b_2(t)z^2 + b_3(t)z^3 + \cdots\},$$

to be the conformal mapping of D onto R_t with

$$g(0,t) = 0, \qquad g'(0,t) = \left.\frac{\partial}{\partial z}g(z,t)\right|_{z=0} = \beta(t) > 0.$$

By Theorem 2.1.6 and the Cauchy Integral Formula, the coefficients of the expansion of g are continuous functions of t. Hence $\beta(t)$ is continuous. Since $g(z,0) = f(z)$, we have $\beta(0) = 1$. By Schwarz lemma, $\beta(t)$ is a strictly increasing function of t. It is now shown that, by reparametrizing Γ, it is possible to make $\beta(t) = e^t$, for $0 \leq t < T$. In general, if $w = \tilde{\psi}(s) = \psi(\sigma(s))$ is another parametrization of Γ, then the corresponding $g'(0,t)$ is given by $\tilde{\beta}(s) = \beta(\sigma(s))$. In particular, if we choose $\sigma(s) = \beta^{-1}(e^s)$, then we obtain $\tilde{\beta}(s) = e^s$.

We now observe that under this parametrization, T must be infinity. For a fixed positive number M, choose t close enough to T so that Γ_t lies entirely outside $|w| = M$. By the Maximum Modulus Theorem

$$\left|\frac{z}{g(z,t)}\right| \leq \frac{1}{M}, \qquad |z| < 1.$$

Then $M \leq |g'(0,t)| = e^t$ for t close enough to T. Since M is arbitrary, we see that $e^t \to \infty$ as $t \to T$. Hence $T = \infty$. Therefore we get the standard parametrization $w = \psi(t)$ of Γ. Under this parametrization,

$$(2.2.1) \qquad g(z,t) = e^t\left\{z + \sum_{n=2}^{\infty} b_n(t)z^n\right\}, \qquad 0 \leq t < \infty.$$

Consider the function

$$(2.2.2) \qquad f(z,t) = g^{-1}(f(z),t) = e^{-t}\left\{z + \sum_{n=2}^{\infty} a_n(t)z^n\right\}, \qquad 0 \leq t < \infty.$$

It maps D onto D minus a slit from a boundary point to its interior. Obviously $f(z,0) = z$ and the coefficients $a_n(t)$ are polynomials of $b_2(t), \cdots, b_n(t)$. Thus the $a_n(t)$ are continuous.

Theorem 2.2.2 (Löwner Differential Equation). *If $f \in S$ is a single slit mapping missing an arc Γ, where $w = \psi(t)$, $0 \leq t < \infty$ is the standard parametrization of Γ, and if $f(z,t)$ is defined by (2.2.2), then $f(z,t)$ satisfies the differential equation*

$$(2.2.3) \qquad \frac{\partial f(z,t)}{\partial t} = -f(z,t)\frac{1 + \kappa(t)f(z,t)}{1 - \kappa(t)f(z,t)},$$

where $\kappa(t)$ is a continuous complex function with $|\kappa(t)| = 1$, $0 \leq t < \infty$, and

$$(2.2.4) \qquad \lim_{t \to \infty} e^t f(z,t) = f(z), \qquad |z| < 1.$$

The convergence is locally uniform.

Equation (2.2.3) is called *Löwner differential equation.*
Proof. We first prove (2.2.4). By Theorem 1.1.4, we have

$$\frac{e^t |z|}{(1 + |z|)^2} \leq |g(z,t)| \leq \frac{e^t |z|}{(1 - |z|)^2}, \qquad |z| < 1.$$

For a fixed $w \in \mathbb{C}$, and a large enough t, let $z = g^{-1}(w,t)$. Then the preceding equation implies that

$$(2.2.5) \qquad \{1 - |g^{-1}(w,t)|\}^2 \leq e^t \left| \frac{g^{-1}(w,t)}{w} \right| \leq \{1 + |g^{-1}(w,t)|\}^2.$$

In particular, we have $|g^{-1}(w,t)| \leq 4|w|e^{-t}$. Hence $g^{-1}(w,t) \to 0$ as $t \to \infty$ uniformly on every compact subset. From (2.2.5), $e^t|g^{-1}(w,t)/w| \to 1$ uniformly on compact subsets. Then $\{e^t g^{-1}(w,t)/w\}_{t \geq 0}$ is a normal family. As t tends to infinity along a sequence of values, $e^t g^{-1}(w,t)/w$ converges to $G(w)$ uniformly on compact subsets. Since $|G(w)| \equiv 1$ and $G(0) = 1$, we get $G(w) \equiv 1$. The limit is independent of the sequence chosen for t, therefore $e^t g^{-1}(w,t)/w \to 1$, as $t \to \infty$. Or equivalently, $e^t g^{-1}(w,t) \to w$ uniformly on compact subsets. This proves (2.2.4)

We now establish Löwner differential equation. For $0 \leq s < t < \infty$, consider the function

$$\zeta = h(z,s,t) = g^{-1}(g(z,s),t) = e^{s-t} z + \cdots .$$

It maps D in the z-plane onto D minus a Jordan curve J_{st} starting from a boundary point in the ζ-plane. Let B_{st} be the part of $|z| = 1$ that corresponds to J_{st}. Let $\lambda(t) = g^{-1}(\psi(t), t)$, the point in the unit circle that is mapped to the end of Γ_t by $g(z,t)$. Then $\lambda(t)$ is the intersecting point between J_{st} and the unit circle. Hence $\lambda(s)$ is an interior point of B_{st}. By the Carathéodory Extension Theorem (If R is a domain bounded by a Jordan curve C, and f maps R conformally onto D, then f can be extended to a homeomorphism of $R \cup C$ onto $\bar{D}$), the function $g^{-1}(w,s)$ is continuous on both sides of Γ_s. Then B_{st} converges to the point $\lambda(s)$ as t decreases to s. Similarly, if t is fixed, then J_{st} converges to point $\lambda(t)$ as s increases to t.

The function $\lambda(t)$ is continuous.

Indeed, applying the Schwarz Reflection Theorem to $h(z, s, t)$ on the complement of B_{st} in the unit circle, we get a function that maps the complement of B_{st} onto the complement of $J_{st} \cup J_{st}^*$, where J_{st}^* is the reflection of J_{st} about the unit circle. By Theorem 1.1.4, the arc J_{st} lies outside the disk $\{\zeta : |\zeta| < \frac{1}{4}e^{s-t}\}$. Its reflection lies inside the disk $\{\zeta : |\zeta| < 4e^{t-\varepsilon}\}$. On the other hand, by the reflection property,

$$\lim_{z \to \infty} \frac{h(z, s, t)}{z} = \lim_{z \to 0} \frac{z}{h(z, s, t)} = e^{t-s}.$$

Hence by the maximum modulus principle, on the complement of B_{st},

$$\left| \frac{h(z, s, t)}{z} \right| \leq 4e^{t-s}.$$

When t decreases to s, B_{st} converges to the point $\lambda(s)$. By the normal family property, we can choose a sequence of t that decreases to s so that the functions $\dfrac{h(z, s, t)}{z}$ converge locally uniformly to an analytic function $\phi(z)$. This function is bounded on the complement of $\lambda(s)$, and satisfies $\phi(0) = 1$. By Liouville Theorem, $\phi(z) \equiv 1$. Since the limit is independent of the choice of the sequence, $h(z, s, t)$ converges to z as t decreases to s, uniformly on every compact set that does not contain $\lambda(s)$.

Fix $s \geq 0$ and $\varepsilon > 0$, and choose $\delta > 0$ small enough so that for $0 < t - s < \delta$, the arc B_{st} is contained in the interior of the circle C centered at $\lambda(s)$ with radius ε. The circle C is mapped into $\tilde{C}$ under the extension of $\zeta = h(z, s, t)$, a Jordan curve that encloses $J_{st} \cup J_{st}^*$. Of course the point $\lambda(t)$ is in the interior of $\tilde{C}$. Since $h(z, s, t)$ converges uniformly to z as $t \to s$, the radius of $\tilde{C}$ will be less than 3ε when t is close enough to s. For each $z_0 \in C$, $t > s$ close enough to s, we have

$$|\lambda(s) - \lambda(t)| \leq |\lambda(s) - z_0| + |z_0 - h(z_0)| + |h(z_0) - \lambda(t)|$$
$$\leq \varepsilon + \varepsilon + 3\varepsilon = 5\varepsilon,$$

where $h(z_0) = h(z_0, s, t)$. Thus λ is right continuous. The left continuity of λ can be proved similarly. Therefore λ is continuous.

Löwner differential equation can then be derived as follows. Let

$$\Phi(z) = \Phi(z, s, t) = \log \left\{ \frac{h(z, s, t)}{z} \right\},$$

where the branch of log is chosen so that $\Phi(0) = s - t$. Then Φ is analytic on D and continuous on $\bar{D}$. From the definition of h, we have $\Re\{\Phi(z)\} = 0$

on the complement of B_{st} in the unit circle, and $\Re e\{\Phi(z)\} < 0$ on B_{st}. By the Schwarz Integral Formula

$$(2.2.6) \qquad \Phi(z) = \frac{1}{2\pi} \int_\alpha^\beta \Re e\{\Phi(e^{i\theta})\} \frac{e^{i\theta} + z}{e^{i\theta} - z}\, d\theta,$$

where $e^{i\alpha}$ and $e^{i\beta}$ are the endpoints of B_{st}. In particular,

$$(2.2.7) \qquad s - t = \Phi(0) = \frac{1}{2\pi} \int_\alpha^\beta \Re e\{\Phi(e^{i\theta})\}\, d\theta.$$

Substituting $f(z, s)$ for z in (2.2.6) and noticing that $h(f(z, s), s, t) = f(z, t)$, we have

$$(2.2.8) \qquad \log \frac{f(z, t)}{f(z, s)} = \frac{1}{2\pi} \int_\alpha^\beta \Re e\{\Phi(e^{i\theta})\} \frac{e^{i\theta} + f(z, s)}{e^{i\theta} + f(z, s)}\, d\theta.$$

The Mean Value Theorem applied to the real and imaginary parts of (2.2.8) yields

$$\log \frac{f(z, t)}{f(z, s)} = \frac{1}{2\pi} \left[\Re e \left\{ \frac{e^{i\sigma} + f(z, s)}{e^{i\sigma} - f(z, s)} \right\} \right.$$
$$\left. + i\ \Im m \left\{ \frac{e^{i\tau} + f(z, s)}{e^{i\tau} - f(z, s)} \right\} \right] \int_\alpha^\beta \Re e\{\Phi(e^{i\theta})\}\, d\theta,$$

where $e^{i\theta}$ and $e^{i\tau}$ are two points on the arc B_{st}. Letting t decrease to s and using (2.2.7), we find that

$$(2.2.9) \qquad \frac{\partial}{\partial s}\{\log f(z, s)\} = -\frac{\lambda(s) + f(z, s)}{\lambda(s) - f(z, s)}$$

because B_{st} converges to the point $\lambda(s)$ as t decreases to s. The derivative in (2.2.9) is the right derivative. Since B_{st} converges to the point $\lambda(t)$ as s increases to t, (2.2.9) also holds for left derivatives. With $\kappa(t) = \dfrac{1}{\lambda(t)}$, (2.2.9) becomes Löwner differential equation (2.2.3). Since λ is continuous and $|\lambda(t)| \equiv 1$, $\kappa(t)$ also has these properties. This completes the proof of the theorem.

If we apply the Mean Value Theorem directly to the integrand of (2.2.6), then

$$\Phi(z) = \frac{1}{2\pi} \left[\Re e \left(\frac{e^{i\sigma} + z}{e^{i\sigma} - z} \right) + i\Im m \left(\frac{e^{i\tau} + z}{e^{i\tau} - z} \right) \right] \int_\alpha^\beta \Re e\{\Phi(e^{i\theta})\}\, d\theta.$$

By (2.2.7), we have

$$\frac{\Phi(z)}{s-t} = \Re e\left(\frac{e^{i\sigma}+z}{e^{i\sigma}-z}\right) + i\Im m\left(\frac{e^{i\tau}+z}{e^{i\tau}-z}\right).$$

However

$$\frac{\Phi(z)}{s-t} = \frac{\log h(z,s,t) - \log z}{s-t}$$

$$= \frac{\log h(z,s,t) - \log z}{g(z,s)-g(z,t)} \cdot \frac{g(z,s)-g(z,t)}{s-t},$$

and $h(z,s,t) \to z$ as $t \to s$. Hence the preceding expression converges to

$$\frac{\partial g(z,s)}{\partial s} \left/ z\frac{\partial g(z,s)}{\partial z} \right.,$$

as $t \to s$. Thus we obtain the differential equation

$$\frac{\partial g(z,t)}{\partial t} = \frac{\partial g(z,t)}{\partial z} z\frac{1+\kappa z}{1-\kappa z}.$$

Theorem 2.2.3 (Alternative form of Löwner Differential Equation). *If $f \in S$ is a single slit mapping missing an arc Γ, $w = \psi(t)$, $0 \le t < \infty$ is a standard parametrization of Γ, and $g(z,t)$ is defined by (2.2.1), then $g(z,t)$ satisfies the differential equation*

$$(2.2.10) \qquad \frac{\partial g(z,t)}{\partial t} = \frac{\partial g(z,t)}{\partial z} z\frac{1+\kappa(t)z}{1-\kappa(t)z},$$

where $\kappa(t)$ is a continuous complex function with $|\kappa(t)| = 1$, $0 \le t < \infty$, and

$$g(z,0) = f(z).$$

Theorem 2.2.3 is the form of Löwner differential equation which de Branges used to prove the Milin Conjecture.

Theorem 2.2.2 states in effect that every single slit mapping $f \in S$ is the locally uniform limit of $e^t f(z,t)$ as $t \to \infty$, where $f(z,t)$ is the solution of Löwner differential equation with continuous unimodular function $\kappa(t)$ and initial condition $f(z,0) = z$. Is the converse also true? Is every solution of Löwner differential equation with continuous unimodular $\kappa(t)$ a single slit mapping? The answer is negative. Kufarev(P.P.Kufarev[1],[2]) established Löwner-Kufarev differential equation to give the more general theorem for the sufficient condition that guarentees the univalency of the solution.

Up to the present time, the necessary and sufficient conditions on κ that will guarantee a slit disk image for $f(z,t)$ have not been found. Only a few partial results have been obtained. For example, Kufarev proved that if $\kappa'(t)$ is continuous on $[0,\infty)$ then $f(z,t)$ maps D into a slit disk. It is also easy to show that if $\kappa(t)$ is constant then the solution $f(z)$ of Löwner's differential equation is a rotation of the Koebe function. If $\kappa(t)$ is a piecewise continuous function that takes only the values ± 1, then $f(z) = z(z^2+cz+1)^{-1}$ where c is a real number and $|c| \leq 2$ (S. Gong [4]). The *Bazilevich functions* arise from the Löwner-Kufarev differential equation for a particular form. Interested readers are referred to I. E. Bazilevich [3] and Duren's book (Duren [1]).

Löwner Method has many applications to univalent function theory. In the next two sections of this chapter, we focus on the applications that relate to the Bieberbach Conjecture. In Chapter 4, we will give a detailed discussion as to how de Branges used Löwner differential equation in proving the Milin Conjecture.

§2.3. The proof of $|a_3| \leq 3$ and related results

Using Löwner method we can prove $|a_3| \leq 3$.

Theorem 2.3.1. *Let $f(z) = z + a_2z^2 + a_3z^3 + \cdots \in S$, then $|a_3| \leq 3$, equality holds if and only if f is the Koebe function or one of its rotations.*

Proof. Since S is invariant under rotations, it is sufficient to prove Re $a_3 \leq 3$. By the Löwner Theory, we only need to prove Re $a_3 \leq 3$ for $f(z) = \lim_{t\to\infty} e^t f(z,t)$, where $f(z,t)$ is the solution of the Löwner differential equation

$$\frac{\partial f}{\partial t} = -f\frac{1+\kappa f}{1-\kappa f}, \qquad f(z,0) = z,$$

where κ is a continuous function with unit modulus. Since

$$f(z,t) = e^{-t}[z + a_2(t)z^2 + a_3(t)z^3 + \cdots],$$

and $a_n(0) = 0$, $\lim_{t\to\infty} a_n(t) = a_n$, $n = 2,3,\ldots$, expanding both sides of Löwner differential equation and equating the coefficients we have

(2.3.1) $a_2'(t) = -2e^{-t}\kappa(t),$

(2.3.2) $a_3'(t) = -2e^{-2t}[\kappa(t)]^2 - 4e^{-t}\kappa(t)a_2(t).$

Integration of (2.3.1) gives

$$(2.3.3) \qquad a_2 = \int_0^\infty a_2'(t)\, dt = -2 \int_0^\infty e^{-t} \kappa(t)\, dt.$$

Since $|\kappa(t)| = 1$, it follows that $|a_2| \leq 2 \int_0^\infty e^{-t}\, dt = 2$, where the equality holds if and only if κ is a constant. But we have seen in §1.1 that this implies that $f(z)$ is a rotation of the Koebe function.

From (2.3.1), we can write (2.3.2) in the following form

$$a_3'(t) = -2e^{-2t}[\kappa(t)]^2 + 2a_2(t)a_2'(t).$$

Integration of this equation gives

$$(2.3.4) \qquad a_3 = -2 \int_0^\infty e^{-2t}[\kappa(t)]^2\, dt + 4\left\{ \int_0^\infty e^{-t}\kappa(t)\, dt \right\}^2.$$

Let $\kappa(t) = e^{i\theta(t)}$, then

$$\Re\{a_3\} \leq 2 \int_0^\infty e^{-2t}[1 - 2\cos^2 \theta(t)]\, dt + 4\left\{ \int_0^\infty e^{-t}\cos\theta(t)\, dt \right\}^2.$$

Applying the Schwarz inequality, we have

$$\Re\{a_3\} \leq 1 - 4 \int_0^\infty e^{-2t}\cos^2\theta(t)\, dt$$

$$+ 4\left\{ \int_0^\infty e^{-t}\, dt \right\}\left\{ \int_0^\infty e^{-t}\cos^2\theta(t)\, dt \right\}$$

$$= 1 + 4 \int_0^\infty (e^{-t} - e^{-2t})\cos^2\theta(t)\, dt$$

$$\leq 1 + 4 \int_0^\infty (e^{-t} - e^{-2t})\, dt = 3.$$

The equality holds if and only if $\cos^2\theta(t) \equiv 1$. Since $\kappa(t)$ is continuous, this is the case only when either $\kappa(t) \equiv 1$ or $\kappa(t) \equiv -1$. This corresponds to the function $z(1+z)^{-2}$ or $z(1-z)^{-2}$.

Thus we have shown that $|a_2| \leq 2$ and $|a_3| \leq 3$ for all $f \in S$ using the Löwner method, and that when the map is single slit, the equality holds if and only if $f(z)$ is a rotation of the Koebe function.

From Theorem 1.1.3, we know that, for any function $f \in S$ defined as (1.1.5), we have $|a_2| \leq 2$. The equality holds if and only if f is the Koebe function defined by (1.1.6) or one of its rotations.

Furthermore, we will now prove that, for any function $f \in S$ defined as (1.1.5), have $|a_3| \leq 3$. The equality holds if and only if f is the Koebe function defined by (1.1.6) or one of its rotations.

This result has been shown to be true when the function f is single slit mapping. By Theorem 2.2.1, for any $f \in S$, there exists a sequence of single slit mappings $f_n \in S$, such that f_n locally uniformly converges to f. Thus $|a_3| \leq 3$ holds for all $f \in S$.

If $|a_3| = 3$ for a function $f \in S$. Without loss of generality, we may assume $a_3 = 3$. For this function f, there exists a sequence of single slit mappings f_n, such that f_n locally uniformly converges to f. Since each f_n is a slit mapping, the second coefficient $a_2(f_n)$ and $a_3(f_n)$ of f_n can be expressed as

$$a_2(f_n) = -2 \int_0^\infty e^{-t} k_n(t) dt$$

and

$$a_3(f_n) = -2 \int_0^\infty e^{-2t} [k_n(t)]^2 dt + 4 \left\{ \int_0^\infty e^{-t} k_n(t) dt \right\}^2$$

by (2.3.3) and (2.3.4), and $a_2(f_n) \to a_2$, $a_3(f_n) \to 3$ as $n \to \infty$.

Let $k_n(t) = e^{i\theta_n(t)}$, then

$$\Re\{a_3(f_n)\}$$

$$= -2 \int_0^\infty e^{-2t} \cos 2\theta_n dt + 4 \left\{ \int_0^\infty e^{-t} \cos \theta_n dt \right\}^2 - 4 \left\{ \int_0^\infty e^{-t} \sin \theta_n dt \right\}^2$$

$$= 1 - 4 \int_0^\infty e^{-2t} \cos^2 \theta_n dt + 4 \left\{ \int_0^\infty e^{-t} \cos \theta_n dt \right\}^2 - 4 \left\{ \int_0^\infty e^{-t} \sin \theta_n dt \right\}^2$$

$$= 3 - 4I_1 - 4I_2 - 4I_3$$

where

$$I_1 = \int_0^\infty (e^{-t} - e^{-2t})(1 - \cos^2 \theta_n) dt,$$

$$I_2 = \int_0^\infty e^{-t} dt \int_0^\infty e^{-t} \cos^2 \theta_n dt - \left\{ \int_0^\infty e^{-t} \cos \theta_n dt \right\}^2$$

and

$$I_3 = \left\{ \int_0^\infty e^{-t} \sin \theta_n dt \right\}^2.$$

Then $I_1 \geq 0$, $I_2 \geq 0$ and $I_3 \geq 0$.

Since $a_3(f_n) \to 3$ is equivalent to $\Re\{a_3(f_n)\} \to 3$, we see that $I_1 \to 0$, $I_2 \to 0$ and $I_3 \to 0$ as $n \to \infty$.

Since $I_3 \to 0$, we have

$$\lim_{n \to \infty} \int_0^\infty e^{-t} \sin \theta_n \, dt = 0.$$

Since $I_1 \to 0$, we have

$$\lim_{n \to \infty} \int_0^\infty (e^{-t} - e^{-2t})(1 - \cos^2 \theta_n) \, dt = 0.$$

Since the integrand is non-negative,

$$\lim_{n \to \infty} \int_c^\infty (e^{-t} - e^{-2t})(1 - \cos^2 \theta_n) \, dt = 0$$

for any $c > 0$. Since $\frac{e^{-t} - e^{-2t}}{1 - e^{-c}} \geq e^{-t}$ for $t > c$, we have

$$\lim_{n \to \infty} \int_c^\infty e^{-t}(1 - \cos^2 \theta_n) \, dt = 0.$$

Obviously

$$\int_0^c e^{-t}(1 - \cos^2 \theta_n) \, dt \leq c.$$

For any positive small number ϵ, we take $c = \frac{\epsilon}{2}$, then we can take a sufficiently large number $N > 0$, such that

$$\int_c^\infty e^{-t}(1 - \cos^2 \theta_n) \, dt < \frac{\epsilon}{2}$$

when $n > N$. Hence we have

$$\int_0^\infty e^{-t}(1 - \cos^2 \theta_n) \, dt < \epsilon$$

when $n > N$. This implies

$$\lim_{n \to \infty} \int_0^\infty e^{-t} \cos^2 \theta_n \, dt = 1.$$

Since $I_2 \to 0$, we have

$$\lim_{n \to \infty} \left\{ \int_0^\infty e^{-t} \cos \theta_n \, dt \right\}^2 = \lim_{n \to \infty} \int_0^\infty e^{-t} \cos^2 \theta_n \, dt = 1.$$

Therefore

$$\lim_{n\to\infty}\left|\int_0^\infty e^{-t}\cos\theta_n\, dt\right| = 1.$$

Using these results to

$$|a_2(f_n)|^2 = \left|-2\int_0^\infty e^{-t}k_n(t)dt\right|^2$$

$$= \left(-2\int_0^\infty e^{-t}\cos\theta_n\, dt\right)^2 + \left(-2\int_0^\infty e^{-t}\sin\theta_n\, dt\right)^2,$$

we have

$$\lim_{n\to\infty}|a_2(f_n)| = 2.$$

Thus

$$\lim_{n\to\infty}|a_2(f_n)| = |a_2| = 2.$$

This proved that: $|a_3| = 3$ if and only if f is the Koebe function or one of its rotations by Theorem 1.1.2.

The case of equality for the Theorem 2.3.1. has not received the attention it deserves.

Löwner in his original paper (K. Löwner[1]) states that equality holds if and only if f is the Koebe function or one of its rotations, but he supplied no proof. The surveys of this work generally fail to even mention the equality case.

The following Feteke-Szegö Theorem is a deep theorem obtained using the Löwner Method.

Theorem 2.3.2 (Feteke-Szegö Theorem). *For every $f \in S$ with expansion (1.1.5), and $0 < \alpha < 1$, the following estimate holds*

(2.3.5) $$|a_3 - \alpha a_2^2| \le 1 + 2e^{-2\alpha/(1-\alpha)}.$$

The estimate is sharp for each fixed α.

The following lemma is used in the proof of the theorem.

Lemma 2.3.1 (Valiron-Landau Lemma). *If $\phi(t)$ is a real continuous function on $t \geq 0$, satisfying the inequalities $|\phi(t)| \leq e^{-t}$ and*

$$(2.3.6) \qquad \int_0^\infty [\phi(t)]^2 \, dt = \left(\lambda + \frac{1}{2}\right) e^{-2\lambda}, \qquad 0 \leq \lambda < \infty,$$

then the following inequality holds

$$\left| \int_0^\infty \phi(t) \, dt \right| \leq (\lambda + 1) e^{-\lambda}.$$

The equality holds if and only if $\phi(t) = \pm\psi(t)$, where

$$\psi(t) = \begin{cases} e^{-\lambda}, & 0 \leq t \leq \lambda, \\ e^{-t}, & \lambda < t < \infty. \end{cases}$$

Proof. Since $|\phi(t)| \leq e^{-t}$, the value of $\int_0^\infty \phi^2 \, dt$ is between 0 and $\frac{1}{2}$. When x increases from 0 to ∞, $\left(x + \frac{1}{2}\right) e^{-2x}$ decreases from $\frac{1}{2}$ to 0. Hence there exists a unique value of λ such that (2.3.6) holds. The function $\psi(t)$ is obviously a continuous function satisfying $|\psi(t)| \leq e^{-t}$ and

$$\int_0^\infty [\psi(t)]^2 \, dt = \left(\lambda + \frac{1}{2}\right) e^{-2\lambda}, \qquad \int_0^\infty \psi(t) \, dt = (\lambda + 1) e^{-\lambda}.$$

Therefore, for all $t \geq 0$, the function

$$F(t) = [\psi(t) - |\phi(t)|][2e^{-\lambda} - \psi(t) - |\phi(t)|]$$

is a non-negative function. Hence

$$0 \leq \int_0^\infty F(t) \, dt = 2e^{-\lambda} \left\{ \int_0^\infty \psi(t) \, dt - \int_0^\infty |\phi(t)| \, dt \right\}$$

$$- \int_0^\infty [\psi(t)]^2 \, dt + \int_0^\infty [\phi(t)]^2 \, dt$$

$$= 2e^{-\lambda} \left\{ (\lambda + 1) e^{-\lambda} - \int_0^\infty |\phi(t)| \, dt \right\},$$

and

$$\left| \int_0^\infty \phi(t) \, dt \right| \leq \int_0^\infty |\phi(t)| \, dt \leq (\lambda + 1) e^{-\lambda}.$$

The equality holds if and only if $\phi(t) = \pm\psi(t)$.

Proof of Theorem 2.3.2. Since the arguments of a_3 and a_2^2 are changed by the same amount under a rotation of $f(z)$, it suffices to consider $\mathrm{Re}\{a_3 - \alpha a_2^2\}$. From (2.3.3) and (2.3.4) we have (setting once again $\kappa(t) = e^{i\theta(t)}$)

$$\Re\{a_3 - \alpha a_2^2\} = 4(1-\alpha)\left\{ \left(\int_0^\infty e^{-t}\cos\theta(t)\,dt \right)^2 \right.$$
$$\left. - \left(\int_0^\infty e^{-t}\sin\theta(t)\,dt \right)^2 \right\}$$
$$- 4 \int_0^\infty e^{-2t}\cos^2\theta(t)\,dt + 1$$
$$\leq 4(1-\alpha)\left\{ \int_0^\infty \phi(t)\,dt \right\}^2 - 4\int_0^\infty [\phi(t)]^2\,dt + 1,$$

where $\phi(t) = e^{-t}\cos\theta(t)$. If

$$\int_0^\infty [\phi(t)]^2\,dt = \left(\lambda + \frac{1}{2} \right)e^{-2\lambda},$$

then Lemma 2.3.1 implies that

$$\Re\{a_3 - \alpha a_2^2\} \leq 4e^{-2\lambda}\left[(1-\alpha)(\lambda+1)^2 - \left(\lambda + \frac{1}{2} \right) \right] + 1.$$

The right hand side of the above inequality attains its maximum value of $1 + 2e^{\frac{-2\lambda}{1-\lambda}}$ when $\lambda = \dfrac{\alpha}{1-\alpha}$. This proves the inequality.

To show that the estimate is sharp, let $\lambda = \dfrac{\alpha}{1-\alpha}$. We only need to show that we can choose $\theta(t)$, $-\frac{\pi}{2} < \theta(t) < \frac{\pi}{2}$ such that $e^{-t}\cos\theta(t) = \psi(t)$. This is done by choosing a piecewise continuous $\theta(t)$ such that

$$\cos\theta(t) = \begin{cases} e^{t-\lambda}, & 0 \leq t \leq \lambda, \\ 1, & \lambda < t < \infty, \end{cases}$$

and

(2.3.7) $$\int_0^\infty e^{-t}\sin\theta(t)\,dt = 0.$$

Choose τ, $0 < \tau < \lambda$ such that $0 < \theta(t) < \frac{\pi}{2}$ for $0 \leq t < \tau$, and that $-\frac{\pi}{2} < \theta(t) \leq 0$ for $\tau \leq t \leq \lambda$. Then

$$\sin\theta(t) = \begin{cases} +\{1 - e^{2(t-\lambda)}\}^{\frac{1}{2}}, & 0 \leq t < \tau, \\ -\{1 - e^{2(t-\lambda)}\}^{\frac{1}{2}}, & \tau \leq t \leq \lambda, \\ 0, & \lambda < t < \infty. \end{cases}$$

Substitute this into the left hand side of (2.3.7). It is easily seen that a suitable τ can be chosen so that (2.3.7) holds.

In §1.3 , it is mentioned that Feteke-Szegö disproved the Littlewood-Paley Conjecture on modulus estimates of coefficients of odd univalent functions. They proved that for an odd univalent function $h(z)$ defined by (1.2.11), $|c_5| \leq \frac{1}{2} + e^{-\frac{2}{3}} = 1.013\cdots$. This is in fact a corollary of Theorem 2.3.2. Since

$$c_5 = \frac{1}{2}\left(a_3 - \frac{1}{4}a_2^2\right),$$

applying (2.3.5) with $\alpha = \frac{1}{4}$ we have $|c_5| \leq \frac{1}{2} + e^{-\frac{2}{3}}$, and this estimate is sharp.

In §1.3, we also mentioned the Robertson Conjecture, and the fact that the Robertson Conjecture implies the Bieberbach Conjecture. We will give a proof of the Robertson Conjecture in the case $n = 3$, using the Löwner method.

Theorem 2.3.3. *If $h(z)$ is a univalent odd function defined by (1.2.11), then*

$$|c_1|^2 + |c_3|^2 + |c_5|^2 \leq 3.$$

Clearly, it can be derived from this result that $|a_3| \leq 3$.
Proof. Since $c_1 = 1$ and we can assume that $c_5 \geq 0$, it suffices to show

$$|c_3|^2 + [\mathrm{Re}\{c_5\}]^2 \leq 2.$$

However we have

$$c_3 = \frac{1}{2}a_2 = -\int_0^\infty e^{-t}\kappa(t)\,dt,$$

$$c_5 = \frac{1}{2}\left(a_3 - \frac{1}{4}a_2^2\right) = \frac{3}{2}\left\{\int_0^\infty e^{-t}\kappa(t)\,dt\right\}^2 - \int_0^\infty e^{-2t}[\kappa(t)]^2\,dt.$$

Putting $\kappa(t) = e^{i\theta(t)}$, $\int_0^\infty e^{-t}\kappa(t)\,dt = u + iv$, then

$$|c_3|^2 = u^2 + v^2,$$

$$c_5 = \mathrm{Re}\{c_5\} = \frac{3}{2}(u^2 - v^2) - 2\int_0^\infty e^{-2t}\cos^2\theta(t)\,dt + \frac{1}{2}.$$

Let $\int_0^\infty e^{-2t} \cos^2 \theta(t)\, dt = \left(\lambda + \frac{1}{2}\right) e^{-2\lambda}$, $0 \le \lambda < \infty$. Then from Lemma 2.3.1, we get $|u| \le (\lambda + 1)e^{-\lambda}$. However since $u^2 + v^2 \le 1$, it follows that

$$|c_3|^2 + |c_5|^2 \le B(\lambda),$$

where

$$B(\lambda) = \min\{1 - v^2, (\lambda + 1)^2 e^{-2\lambda}\} + v^2$$
$$+ \frac{1}{4}\{\beta(\lambda) + 1 - 3v^2\}^2,$$
$$\beta(\lambda) = (3\lambda^2 + 2\lambda + 1)e^{-2\lambda}.$$

It is easy to see that for $0 \le \lambda < \infty$,

$$0 \le \beta(\lambda) \le 2e^{-\frac{2}{3}} = 1.026 \cdots .$$

In the remainder of the proof we treat the following two cases separately:
    Case I. $0 \le 1 - v^2 \le (\lambda + 1)^2 e^{-2\lambda}$. In this case,

$$B(\lambda) = 1 + \frac{1}{4}\{\beta(\lambda) + 1 - 3v^2\}^2.$$

For each fixed λ, this is a quadratic polynomial of v^2 which attains its maximum at the endpoints of the interval. On the end $v^2 = 1 - (\lambda + 1)^2 e^{-2\lambda}$, we have
$$B(\lambda) = 1 + \{(3\lambda^2 + 4\lambda + 2)e^{-2\lambda} - 1\}^2 \le 2,$$
and on the end $v^2 = 1$, we have

$$B(\lambda) = 1 + \frac{1}{4}\{\beta(\lambda) - 2\}^2 \le 2.$$

    Case II. $(\lambda + 1)^2 e^{-2\lambda} \le 1 - v^2 \le 1$. In this case,

$$B(\lambda) = (\lambda + 1)^2 e^{-2\lambda} + v^2 + \frac{1}{4}\{\beta(\lambda) + 1 - 3v^2\}^2.$$

This again is a quadratic polynomial of v^2 which attains its maximum at the endpoints of the interval. On the end $v^2 = 1 - (\lambda + 1)^2 e^{-2\lambda}$, we have already shown that $B(\lambda) \le 2$. On the end $v^2 = 0$, we have

$$B(\lambda) = (\lambda + 1)^2 e^{-2\lambda} + \frac{1}{4}\{\beta(\lambda) + 1\}^2.$$

Since

$$B'(\lambda) = \lambda(1 - 3\lambda)(3\lambda^2 + 2\lambda + 1)e^{-4\lambda} - \lambda(5\lambda + 1)e^{-2\lambda},$$

we have $B'(\lambda) < 0$ for $\frac{1}{3} \leq \lambda < \infty$. For $0 < \lambda < \frac{1}{3}$, using $e^{-4\lambda} < e^{-2\lambda}$, we also have

$$B'(\lambda) < -3\lambda^2(3\lambda^2 + \lambda + 2)e^{-2\lambda} < 0.$$

Hence $B(\lambda) \leq B(0) = 2$ for all $0 \leq \lambda < \infty$. This proves the theorem.

Another theorem concerning coefficients obtained under the Löwner method is the Goluzin Theorem:

Theorem 2.3.4. *Let $f \in S$, with expansion* (1.1.5). *Then*

$$(2.3.8) \qquad -1 \leq |a_3| - |a_2| \leq \frac{3}{4} + e^{-\lambda_0}(2e^{-\lambda_0} - 1) = 1.029\cdots,$$

where λ_0 is the unique solution of $4\lambda e^{-\lambda} = 1$ in the interval $0 < \lambda < 1$. The estimate is sharp.

Proof. Since $f \in S$, by (1.1.20), $|b_1| = |a_2^2 - a_3| \leq 1$. We first prove the left side of (2.3.8). If $|a_2| < 1$, then clearly $|a_2| - |a_3| < 1$. If $|a_2| \geq 1$, then from $|a_2^2 - a_3| \leq 1$, we get

$$|a_2| - |a_3| = |a_2|^2 - |a_3| + |a_2|(1 - |a_2|) \leq 1.$$

Thus the left inequality holds. The equality holds for the starlike function

$$\frac{z}{1 + z + z^2} = z - z^2 + z^3 + \cdots.$$

Now we prove the right side inequality, using the Löwner method. Rotate the function such that $a_3 \geq 0$. We write $a_2 = -2(u + iv)$. Let $\kappa(t) = e^{i\theta(t)}$. Then from (2.3.3) and (2.3.4),

$$|a_3| - |a_2| = \operatorname{Re}\{a_3\} - |a_2| = -4\int_0^\infty e^{-2t}\cos^2\theta(t)\,dt$$
$$+ 1 + 4(u^2 - v^2) - 2\{u^2 + v^2\}^{\frac{1}{2}}$$
$$\leq 1 + 4u^2 - 2|u| - 4\int_0^\infty e^{-2t}\cos^2\theta(t)\,dt.$$

Let the last integral be $\left(\lambda + \frac{1}{2}\right)e^{-2\lambda}$, $0 \leq \lambda < \infty$. By Lemma 2.3.1, $|u| \leq (\lambda + 1)e^{-\lambda}$. We consider two cases.

Case I. $(\lambda + 1)e^{-\lambda} \leq \frac{1}{2}$. In this case $|u| \leq \frac{1}{2}$, $4u^2 - 2|u| \leq 0$, hence $|a_3| - |a_2| \leq 1$.

Case II. $(\lambda + 1)e^{-\lambda} > \frac{1}{2}$. In this case, $4u^2 - 2|u|$ is positive and is increasing in $|u|$ over the interval $\frac{1}{2} < |u| < \infty$. Then

$$|a_3| - |a_2| \leq 1 + 4(\lambda + 1)^2 e^{-2\lambda} - 2(\lambda + 1)e^{-\lambda} - 4\left(\lambda + \frac{1}{2}\right)e^{-2\lambda}$$

$$= 1 + 2(2\lambda^2 + 2\lambda + 1)e^{-2\lambda} - 2(\lambda + 1)e^{-\lambda}.$$

Hence the problem is reduced to finding the maximum of the function

$$\phi(\lambda) = (2\lambda^2 + 2\lambda + 1)e^{-2\lambda} - (\lambda + 1)e^{-\lambda}$$

over the given interval. However

$$\phi'(\lambda) = \lambda e^{-\lambda}(1 - 4\lambda e^{-\lambda}),$$

hence $\phi(\lambda)$ takes its extremum at the points $\lambda = 0$, λ_0, λ_1, where $0 < \lambda_0 < 1$, and $\lambda_1 > 2$ are the roots of the equation $4\lambda e^{-\lambda} = 1$. Since $(\lambda_1 + 1)e^{-\lambda_1} < \frac{1}{2}$, we drop λ_1. The second derivative of $\phi(\lambda)$ is

$$\phi''(\lambda) = (1 - \lambda)e^{-\lambda}(1 - 4\lambda e^{-\lambda}) + 4\lambda(\lambda - 1)e^{-2\lambda}.$$

Therefore $\phi''(\lambda_0) = (\lambda_0 - 1)e^{-\lambda_0} < 0$, and λ_0 is a local maximum point of ϕ. Numerical calculation gives us $\lambda_0 = 0.35740 \cdots$. Noticing that $\lambda_0 e^{-\lambda_0} = \frac{1}{4}$, we get

$$|a_3| - |a_2| \leq \frac{3}{4} + e^{-\lambda_0}(2e^{-\lambda_0} - 1) = 1.029 \cdots.$$

An argument similar to that in Theorem 2.3.2 can be used to prove the sharpness of the estimate.

The estimates of difference of modulus between general successive coefficients will be given in §3.4.

The three theorems of this section share the same scheme of proof. They all start from (2.3.3) and (2.3.4), and then make use of Lemma 2.3.1. The equations (2.3.3) and (2.3.4) are obtained by expanding Löwner differential equation and comparing coefficients. The expression for a_n obtained this way is very complicated for large n's, making it nearly impossible to prove $|a_n| \leq n$ using only these expressions. Although we proved $|a_3| \leq 3$ from

(2.3.4), the proof of $|a_4| \leq 4$ from the corresponding expression is highly non-trivial. The estimate $|a_3| \leq 3$ was proved by Löwner by using his method in 1923. And fifty years had passed before a proof of $|a_4| \leq 4$ using Löwner method was obtained by Nehari (Z. Nehari [2]). We will not discuss Nehari's proof in this book.

By the way, F. R. Keogh and E. P. Merkes [1] proved the corresponding results of Theorem 2.3.2 for convex functions, starlike functions and close-to-convex functions.

§2.4. FitzGerald Inequality

In 1972, FitzGerald exponentiated the Goluzin Inequality (C. H. FitzGerald [1], [2], C. H. FitzGerald and R. A. Horn [1]), and obtained the FitzGerald Inequality. He proved several coefficient inequalities, and the following important result: if $f \in S$ with expansion (1.1.5), then $|a_n| < \sqrt{\frac{7}{6}}\, n < 1.081n$ for all n. Later, Horowitz used FitzGerald inequality to improve this result to $|a_n| < \left(\frac{209}{140}\right)^{\frac{1}{6}} n < 1.0691n$ and again to show that $|a_n| < 1.0657n$ (D.Horowitz [1,2]). In proving his inequality, FitzGerald used the Löwner method. In this section, we give his proof.

We prove the Goluzin Inequality (G. M. Goluzin [3,4,5]) first.

We will restrict our discussion to Löwner Functions, that is, functions $f(z) \in S$ for which there exists a function $f(z,t)$ that is analytic and univalent in $|z| < 1$, with $|f(z,t)| < 1$, $f(0,t) = 0$, $f'(0,t) > 0$, satisfies equation (2.2.3), and

$$\lim_{t \to \infty} e^t f(z,t) = f(z).$$

Let $\zeta = \dfrac{1}{z}$, $F(\zeta) = 1 \Big/ f\left(\dfrac{1}{z}\right) \in \Sigma$, $z_1, \ldots, z_n$ be n arbitrary points in $|z| < 1$, and $\zeta_\nu = \dfrac{1}{z_\nu}$, $f_\nu = f(z_\nu, t)$. Then direct computation from

$$\frac{\partial f_\nu}{\partial t} = -f_\nu \frac{1 + \kappa(t) f_\nu}{1 - \kappa(t) f_\nu}, \qquad \nu = 1, \ldots, n,$$

yields

$$(2.4.1) \qquad \frac{\partial}{\partial t} \log\left[\frac{e^{-t}}{f_\nu f_\mu} \cdot \frac{f_\nu - f_\mu}{z_\nu - z_\mu}\right] = -2\frac{\kappa f_\nu}{1 - \kappa f_\nu} \cdot \frac{\kappa f_\mu}{1 - \kappa f_\mu},$$

$$(2.4.2) \qquad \frac{\partial}{\partial t} \log(1 - f_\nu \bar{f}_\mu) = 2\frac{\kappa f_\nu}{1 - \kappa f_\nu}\left(\overline{\frac{\kappa f_\mu}{1 - \kappa f_\mu}}\right).$$

Integrating the preceding two equations from 0 to ∞, and observing $f(z,0) = z$, we obtain

(2.4.3)
$$\log \frac{F(\zeta_\nu) - F(\zeta_\mu)}{\zeta_\nu - \zeta_\mu} = -2 \int_0^\infty \frac{\kappa f_\nu}{1 - \kappa f_\nu} \cdot \frac{\kappa f_\mu}{1 - \kappa f_\mu} \, dt,$$

(2.4.4)
$$\log \left(1 - \frac{1}{\zeta_\nu \bar{\zeta}_\mu} \right) = -2 \int_0^\infty \frac{\kappa f_\nu}{1 - \kappa f_\nu} \left(\overline{\frac{\kappa f_\mu}{1 - \kappa f_\mu}} \right) dt.$$

Let $\dfrac{\kappa f_\nu}{1 - \kappa f_\nu} = X_\nu + i Y_\nu$, $\nu = 1, \ldots, n$. Taking the real parts of (2.4.3) and (2.4.4), we find

$$\log \left| \frac{F(\zeta_\nu) - F(\zeta_\mu)}{\zeta_\nu - \zeta_\mu} \right| = -2 \int_0^\infty (X_\nu X_\mu - Y_\nu Y_\mu) \, dt,$$

$$\log \left| 1 - \frac{1}{\zeta_\nu \bar{\zeta}_\mu} \right| = -2 \int_0^\infty (X_\nu X_\mu + Y_\nu Y_\mu) \, dt.$$

Hence

(2.4.5)
$$\log \left| \frac{F(\zeta_\nu) - F(\zeta_\mu)}{\zeta_\nu - \zeta_\mu} \right| = -\log \left| 1 - \frac{1}{\zeta_\nu \bar{\zeta}_\mu} \right| - 4 \int_0^\infty X_\nu X_\mu \, dt,$$

(2.4.6)
$$\log \left| \frac{F(\zeta_\nu) - F(\zeta_\mu)}{\zeta_\nu - \zeta_\mu} \right| = \log \left| 1 - \frac{1}{\zeta_\nu \bar{\zeta}_\mu} \right| + 4 \int_0^\infty Y_\nu Y_\mu \, dt.$$

From (2.4.5) and (2.4.6) for any positive definite $n \times n$ matrix $A = (a_{\mu\nu})$,

(2.4.7)
$$\sum_{\mu,\nu=1}^n a_{\mu\nu} \log \left| 1 - \frac{1}{\zeta_\mu \bar{\zeta}_\nu} \right| \leq \sum_{\mu,\nu=1}^n a_{\mu\nu} \log \left| \frac{F(\zeta_\mu) - F(\zeta_\nu)}{\zeta_\mu - \zeta_\nu} \right|$$
$$\leq -\sum_{\mu,\nu=1}^n a_{\mu\nu} \log \left| 1 - \frac{1}{\zeta_\mu \bar{\zeta}_\nu} \right|.$$

This is one of the Goluzin Inequalities.

From (2.4.3),

$$\sum_{\mu,\nu=1}^n \gamma_\mu \gamma_\nu \log \frac{F(\zeta_\mu) - F(\zeta_\nu)}{\zeta_\mu - \zeta_\nu}$$
$$= -2 \int_0^\infty \sum_{\mu,\nu=1}^n \gamma_\mu \gamma_\nu \frac{\kappa f_\mu}{1 - \kappa f_\mu} \cdot \frac{\kappa f_\nu}{1 - \kappa f_\nu} \, dt$$
$$= -2 \int_0^\infty \left(\sum_{\nu=1}^n \gamma_\nu \frac{\kappa f_\nu}{1 - \kappa f_\nu} \right)^2 dt.$$

Consider the absolute value of this equation,

$$\left| \sum_{\mu,\nu=1}^{n} \gamma_\mu \gamma_\nu \log \frac{F(\zeta_\mu) - F(\zeta_\nu)}{\zeta_\mu - \zeta_\nu} \right| \leq 2 \int_0^\infty \left| \sum_{\nu=1}^{n} \gamma_\nu \frac{\kappa f_\nu}{1 - \kappa f_\nu} \right|^2 dt$$

$$= 2 \int_0^\infty \sum_{\mu,\nu=1}^{n} \gamma_\mu \bar{\gamma}_\nu \frac{\kappa f_\nu}{1 - \kappa f_\nu} \overline{\left(\frac{\kappa f_\nu}{1 - \kappa f_\nu} \right)} dt.$$

By (2.4.4), the right hand side of the preceding inequality equals

$$-\sum_{\mu,\nu=1}^{n} \gamma_\mu \bar{\gamma}_\nu \log \left(1 - \frac{1}{\zeta_\mu \bar{\zeta}_\nu} \right).$$

Hence we have proved another one of the Goluzin Inequalities:

$$(2.4.8) \qquad \left| \sum_{\mu,\nu=1}^{n} \gamma_\mu \gamma_\nu \log \frac{F(\zeta_\mu) - F(\zeta_\nu)}{\zeta_\mu - \zeta_\nu} \right| \leq -\sum_{\mu,\nu=1}^{n} \gamma_\mu \bar{\gamma}_\nu \log \left(1 - \frac{1}{\zeta_\mu \bar{\zeta}_\nu} \right).$$

Some applications of these two inequalities are presented in Goluzin's book (G.M. Goluzin [1]).

FitzGerald "exponentiated" the Goluzin Inequalities by dropping the log from the inequalities.

From (2.4.6), we have

$$4 \int_0^\infty Y_\mu Y_\nu \, dt = \log \left| \frac{F(\zeta_\mu) - F(\zeta_\nu)}{\zeta_\mu - \zeta_\nu} \cdot \frac{\zeta_\mu \bar{\zeta}_\nu}{\zeta_\mu \zeta_\nu - 1} \right|.$$

Thus

$$0 \leq \int_0^\infty \cdots \int_0^\infty \left| \sum_{\mu=1}^{n} \beta_\mu Y_\mu(t_1) \cdots Y_\mu(t_m) \right|^2 dt_1 \cdots dt_m$$

$$= \sum_{\mu,\nu=1}^{n} \beta_\mu \bar{\beta}_\nu \left(\int_0^\infty Y_\mu(t) Y_\nu(t) \, dt \right)^m$$

$$= \sum_{\mu,\nu=1}^{n} \beta_\mu \bar{\beta}_\nu \left(\frac{1}{4} \log \left| \frac{(F(\zeta_\mu) - F(\zeta_\nu))\zeta_\mu \bar{\zeta}_\nu}{(\zeta_\mu - \zeta_\nu)(\zeta_\mu \bar{\zeta}_\nu - 1)} \right| \right)^m.$$

Multiplying both sides of the above inequality by $\dfrac{8^m}{m!}$ and then summing over 1 to ∞, we get

$$\sum_{\mu,\nu=1}^{n} \beta_\mu \bar{\beta}_\nu \left(\left| \frac{(F(\zeta_\mu) - F(\zeta_\nu))\zeta_\mu \bar{\zeta}_\nu}{(\zeta_\mu - \zeta_\nu)(\zeta_\mu \bar{\zeta}_\nu - 1)} \right|^2 - 1 \right) \geq 0,$$

or

$$\sum_{\mu=1}^{n}\sum_{\nu=1}^{n}\beta_{\mu}\bar{\beta}_{\nu}\left(\left|\frac{f(z_{\mu})-f(z_{\nu})}{f(z_{\mu})\cdot f(z_{\nu})}\frac{z_{\mu}\bar{z}_{\nu}}{(z_{\mu}-z_{\nu})(1-z_{\mu}\bar{z}_{\nu})}\right|^{2}-1\right)\geq 0.$$

Letting $\beta_{\nu}=\left|\dfrac{f(z_{\nu})}{z_{\nu}}\right|^{2}\alpha_{\nu}$, we arrive at FitzGerald Inequality.

Theorem 2.4.1. *If $f(z)\in S$, $\alpha_{1},\ldots,\alpha_{n}$ are n arbitrary complex numbers, and $z_{1},\ldots,z_{n}$ are n arbitrary points in $|z|<1$, then*

(2.4.9)
$$\sum_{\mu=1}^{n}\sum_{\nu=1}^{n}\alpha_{\mu}\bar{\alpha}_{\nu}\left|\frac{f(z_{\mu})-f(z_{\nu})}{z_{\mu}-z_{\nu}}\cdot\frac{1}{1-z_{\mu}\bar{z}_{\nu}}\right|^{2}$$
$$\geq\sum_{\mu=1}^{n}\sum_{\nu=1}^{n}\alpha_{\mu}\bar{\alpha}_{\nu}\left|\frac{f(z_{\mu})f(z_{\nu})}{z_{\mu}z_{\nu}}\right|^{2}.$$

From Theorem 2.4.1, we have the following FitzGerald Coefficient Inequality.

Theorem 2.4.2 (FitzGerald Inequality). *Let $f(z)=z+\sum_{n=2}^{\infty}a_{n}z^{n}\in S$. Define $\beta_{n}(\mu,\nu)$ as follows: for $\mu\leq\nu$,*

(2.4.10)
$$\beta_{n}(\mu,\nu)=\begin{cases}\mu-|n-\nu|, & |n-\nu|<\mu,\\ 0, & |n-\nu|\geq\mu,\end{cases}$$

and $\beta_{n}(\nu,\mu)=\beta_{n}(\mu,\nu)$. Denote

(2.4.11)
$$a_{\mu\nu}(f)=\sum_{k=1}^{\mu+\nu-1}\beta_{k}(\mu,\nu)|a_{k}|^{2}-|a_{\mu}|^{2}|a_{\nu}|^{2},$$

then the matrix

(2.4.12)
$$(a_{\mu\nu})_{2\leq\mu,\nu\leq n}\geq 0.$$

Proof. Set $n=Nm$ in (2.4.9). Let z_{ν}, α_{ν} be given by $z_{p\mu}=r_{p}e^{2\pi i\mu/m}$, $\alpha_{p\mu}=m^{-1}\gamma_{p}$ respectively, where $0<r_{p}<1$. Let $m\to\infty$, then the sum in (2.4.9) becomes an integral, and

(2.4.13)
$$\left(\sum_{p=1}^{N}\frac{\gamma_{p}}{2\pi}\int_{0}^{2\pi}\left|\frac{f(r_{p}e^{it})}{r_{p}e^{it}}\right|^{2}dt\right)^{2}$$
$$\leq\sum_{p=1}^{N}\sum_{q=1}^{N}\frac{\gamma_{p}\gamma_{q}}{(2\pi)^{2}}\int_{0}^{2\pi}\int_{0}^{2\pi}\left|\frac{f(r_{p}e^{is})-f(r_{q}e^{it})}{(r_{p}e^{is}-r_{q}e^{it})(1-r_{p}r_{q}e^{i(s-t)})}\right|^{2}ds\,dt.$$

Now we compute the integral in the above equation. Since

$$\frac{f(z)-f(\zeta)}{z-\zeta} = \sum_{k=0}^{\infty}\sum_{\ell=0}^{\infty} a_{k+\ell+1} z^k \zeta^\ell, \qquad \frac{1}{1-z\bar{\zeta}} = \sum_{j=0}^{\infty} z^j \bar{\zeta}^j,$$

the integrand in the right hand side is

$$\sum_{k,\ell,j\geq 0}\sum_{k',\ell',j'\geq 0} a_{k+\ell+1}\bar{a}_{k'+\ell'+1} r_p^{k+j+k'+j'} r_q^{\ell+j+\ell'+j'} e^{i(k+j-k'-j')s}$$
$$\times e^{i(\ell-j-\ell'+j')t}.$$

The integral of all terms except terms with $k+j = k'+j'$, $\ell - j = \ell' - j'$ are zero. Set

$$n = k + \ell + 1 = k' + \ell' + 1,$$
$$\mu = k + j + 1 = k' + j' + 1,$$
$$\nu = \ell + j' + 1 = \ell' + j + 1.$$

Then the integral on the right hand side of (2.4.13) becomes

$$\sum_{\mu=1}^{\infty}\sum_{\nu=1}^{\infty}\sum_{n=1}^{\infty} |a_n|^2 r_p^{2(\mu-1)} r_q^{2(\nu-1)} \beta_n(\mu,\nu).$$

Here $\beta_n(\mu,\nu)$ is the number of integers j that satisfies the following inequalities:

$$0 \leq j \leq \mu + \nu - n - 1,$$
$$\mu - n \leq j \leq \mu - 1,$$
$$\nu - n \leq j \leq \nu - 1.$$

When $\mu \leq \nu$, these conditions become: when $n \leq \nu$, $\nu - n < j \leq \mu - 1$; when $n > \nu$, $0 \leq j \leq \mu + \nu - n - 1$. But this is exactly (2.4.10).

Then the integral on the left hand side of (2.4.13) equals

$$\sum_{n=1}^{\infty} |a_n|^2 r_p^{2(n-1)}.$$

Thus (2.4.13) becomes

(2.4.13')
$$\left(\sum_{n=1}^{\infty} |a_n|^2 \lambda_n\right)^2 \leq \sum_{\mu=1}^{\infty}\sum_{\nu=1}^{\infty}\sum_{n=1}^{\infty} |a_n|^2 \lambda_\mu \lambda_\nu \beta_n(\mu,\nu),$$

where $\lambda_\nu = \sum_{p=1}^{N} \gamma_p r_p^{2(\nu-1)}$, $\nu = 1, 2, \ldots$.

Let $0 < s_1 < \cdots < s_N < 1$. Then for any $n = 1, \ldots, N$, we can choose real numbers α_{np}, $p = 1, \ldots, N$, so that for any real numbers $x_1, \ldots, x_N$,

$$(2.4.14) \qquad \sum_{p=1}^{N} s_p^{2(p-1)} \alpha_{np} = \begin{cases} x_n, & \nu = n; \\ 0, & \nu = 1, \ldots, N, \nu \neq n. \end{cases}$$

The above system of equations has a solution because the coefficients determinant of α_{np} is a Vandermonde determinant. Choose $0 < \delta < 1$, and let

$$r_p = \delta s_p, \qquad \gamma_p = \sum_{n=1}^{N} \delta^{-2(n-1)} \alpha_{np}, \qquad p = 1, \ldots, N.$$

Then

$$\lambda_\nu = \lambda_\nu(\delta) = \sum_{n=1}^{N} \delta^{2(\nu-n)} \sum_{p=1}^{N} s_p^{2(\nu-1)} \alpha_{np}, \qquad \nu = 1, 2, \ldots.$$

From (2.4.14), for $1 \leq \nu \leq N$, we have $\lambda_\nu = x_\nu$; and from the above equation, for $\nu > N$, we have $\lambda_n = O(\delta^{2(\nu-N)})$. Letting $\delta \to 0$, we get

$$(2.4.12') \qquad \left(\sum_{n=1}^{N} |a_n|^2 x_n \right)^2 \leq \sum_{\mu=1}^{N} \sum_{\nu=1}^{N} \sum_{n=1}^{\mu+\nu-1} \beta_n(\mu, \nu) |a_n|^2 x_\mu x_\nu$$

for any real numbers $x_1, \ldots, x_n$. This is (2.4.12).

The main diagonal entries of (2.4.12) are non-negative. We have the following corollary.

Corollary 2.4.1. *If $f(z) = z + \sum_{n=2}^{\infty} a_n z^n \in S$, then*

$$(2.4.15) \qquad |a_n|^4 \leq \sum_{k=1}^{n} k |a_k|^2 + \sum_{k=n+1}^{2n-1} (2n-k) |a_k|^2, \qquad n = 2, 3, \ldots.$$

From this corollary, we get the following corollary.

Corollary 2.4.3. *If $f(z) = z + \sum_{n=2}^{\infty} a_n z^n \in S$, then*

$$|a_n| < \sqrt{\frac{7}{6}}\, n < 1.081n, \qquad n = 2, 3, \ldots .$$

Proof. Since

$$\sum_{k=1}^{n} k^2 = \frac{1}{6}n(n+1)(2n+1), \qquad \sum_{k=1}^{n} k^3 = \frac{1}{4}n^2(n+1)^2,$$

under the substitution of $|a_k|$ by k, the right hand side of (2.4.15) becomes

$$\sum_{k=1}^{n} k^3 + \sum_{k=n+1}^{2n-1} (2n-k)k^2 = \frac{7}{6}n^4 - \frac{1}{6}n^2.$$

Let $c = \sup_n \sup_{f \in S} \dfrac{|a_n|}{n}$, i.e., let c be the smallest constant such that for all $f \in S$ and all n, $|a_n| \leq cn$. By Theorem 1.2.3 (Littlewood Theorem), we have $1 \leq c \leq e$. For any $\varepsilon > 0$, there exists an integer n and an $f \in S$, such that $|a_n| \geq (c - \varepsilon)n$. Thus by (2.4.15), we have

$$(c-\varepsilon)^4 n^4 < |a_n|^4 \leq c^2\left(\frac{7}{6}n^4 - \frac{1}{6}n^2\right).$$

Consequently, $(c-\varepsilon)^4 < \frac{7}{6}c^2 - \frac{1}{6}n^2$. Let $\varepsilon \to 0$, we get $c < \sqrt{\frac{7}{6}}$.

In the proof of Corollary 2.4.3, we only used Corollary 2.4.2. But Corollary 2.4.2 is only a simple consequence of $(a_{\mu\nu})_{2 \leq \mu, \nu \leq n} \geq 0$, i.e., its main diagonal entries are non-negative. From this point of view, the result of Corollary 2.4.3 can be improved. Horowitz (D. Horowitz [1, 2]) obtained this kind of improvement. He started from $(u_{\mu\nu})_{2 \leq \mu, \nu \leq 2n} \geq 0$, and considered $\lambda = (\lambda_2, \ldots, \lambda_{2n})$, where $\lambda_\mu = n - |n - \mu|$, $\mu = 1, 2, \ldots, 2n$. The inequality $\lambda(a_{\mu\nu})\lambda' \geq 0$ can be written as

$$\left|\sum_{\mu=1}^{2n} \lambda_\mu |a_\mu|^2\right|^2 \leq \sum_{\mu=1}^{2n} |\lambda_\mu|^2 \left\{\sum_{k=1}^{\mu} k|a_k|^2 + \sum_{k=\mu+1}^{2\mu} (2\mu - k)|a_k|^2\right\}$$

$$+ 2 \sum_{1 \leq \mu_1 < \mu_2 \leq 2n} \lambda_{\mu_1}\lambda_{\mu_2}$$

$$\times \left\{= \sum_{k\mu_2 - \mu_1}^{\mu_2} (\mu_1 - \mu_2 + k)|a_k|^2 + \sum_{k=\mu_2+1}^{\mu_1+\mu_2} (\mu_1 + \mu_2 - k)|a_k|^2\right\}.$$

The left hand side of this inequality is the square of the right hand side of (2.4.15). Hence

$$|a_n|^8 \leq \sum_{\mu=1}^{2n} \lambda_\mu^2 \left\{ \sum_{k=1}^{\mu} k|a_k|^2 + \sum_{k=\mu+1}^{2\mu} (2\mu - k)|a_k|^2 \right\}$$

(2.4.16)

$$+ 2 \sum_{\ell=2}^{2n} \sum_{m=1}^{\ell-1} \lambda_\ell \lambda_m \left\{ \sum_{k=\ell-m}^{\ell} (m - \ell + k)|a_k|^2 \right.$$

$$\left. + \sum_{k=\ell+1}^{\ell+m} (m + \ell - k)|a_k|^2 \right\}.$$

Again let $c = \sup_n \sup_{f \in S} \dfrac{|a_n|}{n}$. Then for any $\varepsilon > 0$, there exists n and $f(z) \in S$ such that $|a_n| > n(c - \varepsilon)$. The same argument as in the proof of Corollary 2.4.3 leads to

$$n^8 (c - \varepsilon)^8 \leq c^2 \left[\sum_{\mu=1}^{2n} \lambda_\mu^2 \left\{ \sum_{k=1}^{\mu} k^3 + \sum_{k=\mu+1}^{2\mu} (2\mu - k)k^2 \right\} \right.$$

$$\left. + 2 \sum_{\ell=2}^{2n} \sum_{m=1}^{\ell-1} \lambda_\ell \lambda_m \left\{ \sum_{k=\ell-m}^{\ell} (m - \ell + k)k^2 + \sum_{k=\ell+1}^{\ell+m} (m + \ell - k)k^2 \right\} \right].$$

Making use of the summation formulas for $\sum_{k=1}^{N} k^\ell$, $\ell = 1, 2, \ldots, 7$, we can see through a lengthy calculation that the right hand side of the above inequality is equal to

$$c^2 \left[\frac{1}{1260} (1881n^8 - 602n^6 + 49n^4 - 68n^2) \right] < c^2 \frac{1881}{1260} n^8.$$

Since ε is arbitrary, we let $\varepsilon \to 0$ to obtain $c^6 \leq \frac{209}{140}$, i.e., $c \leq \left(\frac{209}{140} \right)^{\frac{1}{6}} <$ 1.0691. We get the result: If $f \in S$, then $|a_n| \leq \left(\frac{209}{140} \right)^{\frac{1}{6}} n < 1.0691n$, $n = 2, 3, \ldots$. This result was obtained by Horowitz in 1976. In 1978, he improved it to $|a_n| \leq \left(\frac{1659164137}{681080400} \right)^{\frac{1}{14}} n < 1.0657n$, $n = 2, 3, \ldots$. by using FitzGerald inequality again and again.

The Bieberbach Conjecture, however, cannot be proved this way. The result $|a_n| < 1.0657n$ is the best result on the general estimates of the coefficients before de Branges' proof of the Bieberbach Conjecture.

CHAPTER III

GRUNSKY INEQUALITY

§3.1. Faber Polynomials, Grunsky Inequality

Let $g(\zeta) = \zeta + b_0 + b_1\zeta^{-1} + b_2\zeta^{-2} + \cdots \in \Sigma$. Consider the expansion

(3.1.1)
$$\frac{\zeta g'(\zeta)}{g(\zeta) - w} = \sum_{n=0}^{\infty} F_n(w)\zeta^{-n},$$

at a neighborhood of $\zeta = \infty$ where $w \in \mathbb{C}$. The function

$$F_n(w) = w^n + \sum_{k=1}^{n} a_{nk} w^{n-k}$$

is an n-th degree polynomial in w, called the *Faber polynomial* of the function g. Direct computation shows that

$$
\begin{aligned}
F_0(w) &= 1, \\
F_1(w) &= w - b_0, \\
F_2(w) &= (w - b_0)^2 - 2b_1 = w^2 - 2b_0 w + (b_0^2 - 2b_1), \\
F_3(w) &= (w - b_0)^3 - 3b_1(w - b_0) - 3b_1 \\
&= w^3 - 3b_0 w^2 + (3b_0^2 - 3b_1)w + (b_0^3 + 3b_1 b_0 - 3b_2), \\
F_4(w) &= (w - b_0)^4 - 4b_1(w - b_0)^2 - 4b_2(w - b_0) + (2b_1^2 - 4b_3) \\
&= w^4 - 4b_0 w^3 + (6b_0^2 - 4b_1)w^2 + (-4b_0^3 + 8b_0 b_1 - 4b_2)w \\
&\quad + (b_0^4 - 4b_0^2 b_1 + 4b_0 b_2 + 2b_1^2 - 4b_3), \\
&\cdots\cdots .
\end{aligned}
$$

Since g is univalent, the function

(3.1.2) $$\frac{\zeta g'(\zeta)}{g(\zeta) - g(z)} - \frac{\zeta}{\zeta - z} = \sum_{n=1}^{\infty} \sum_{k=1}^{\infty} \beta_{nk} z^{-k} \zeta^{-n}$$

is holomorphic in $|z| > 1$, $|\zeta| > 1$. By (3.1.1), we have

$$\sum_{n=0}^{\infty} F_n(g(z)) \zeta^{-n} = 1 + \sum_{n=1}^{\infty} \left\{ z^n + \sum_{k=1}^{\infty} \beta_{nk} z^{-k} \right\} \zeta^{-n}.$$

The Faber polynomials satisfy

(3.1.3) $$F_n(g(z)) = z^n + \sum_{k=1}^{\infty} \beta_{nk} z^{-k}, \qquad n = 1, 2, \ldots.$$

The coefficients β_{nk} are called the *Grunsky coefficients* of g.

We mentioned in §1.2 (1.2.1) that if $g \in \Sigma$, then

$$\log \frac{g(\zeta) - g(z)}{\zeta - z} = -\sum_{n=1}^{\infty} \sum_{k=1}^{\infty} \gamma_{nk} z^{-k} \zeta^{-n}$$

is holomorphic in $\{(z, \zeta) : |z| > 1, |\zeta| > 1\}$. This function is symmetric in z and ζ, hence $\gamma_{nk} = \gamma_{kn}$. Dividing (3.1.2) by ζ and then integrating with respect to ζ, we obtain

$$\log \frac{g(\zeta) - g(z)}{\zeta - z} = -\sum_{n=1}^{\infty} \sum_{k=1}^{\infty} \frac{1}{n} \beta_{nk} z^{-k} \zeta^{-n}.$$

It follows that $\beta_{nk} = n\gamma_{nk}$, $k\beta_{nk} = nk\gamma_{nk}$. Hence $k\beta_{nk} = n\beta_{kn}$, $k, n = 1, 2, \ldots$.

By direct computation, we have

$$\beta_{11} = b_1, \quad \beta_{12} = b_2, \quad \beta_{13} = b_3, \quad \beta_{14} = b_4, \quad \cdots$$

$$\beta_{21} = 2b_2, \quad \beta_{22} = 2b_3 + b_1^2, \quad \beta_{23} = 2(b_4 + b_1 b_2),$$

$$\beta_{24} = 2b_5 + 2b_1 b_3 + b_2^2, \quad \cdots$$

$$\beta_{31} = 3b_3, \quad \beta_{32} = 3(b_4 + b_1 b_2), \quad \beta_{33} = 3(b_5 + b_1 b_3 + b_2^2) + b_1^3,$$

$$\beta_{34} = 3(b_6 + b_1 b_4 + 2b_2 b_3 + b_1^2 b_2), \quad \cdots$$

$$\beta_{41} = 4b_4, \quad \beta_{42} = 4(b_5 + b_1 b_3) + 2b_2^2,$$

$$\beta_{43} = 4(b_6 + b_1 b_4 + 2b_2 b_3 + b_1^2 b_2),$$

$$\beta_{44} = 4(b_7 + b_1 b_5 + b_1^2 b_3) + 8(b_2 b_4 + b_1 b_2^2) + 6b_3^2 + b_1^4, \quad \cdots$$

$$\cdots \cdots$$

In §1.2 we stated Grunsky Inequality (1.2.2). There are another two equivalent forms of Grunsky Inequality, the Strong Grunsky Inequality, (1.2.3) and the Generalized Weak Grunsky Inequality (1.2.4). We give the proofs of these inequalities now. (Grunsky [1]).

Theorem 3.1.1 (Grunsky Inequality). *Let β_{nk} be the Grunsky coefficients of g. Then for any positive integer N and complex numbers $\lambda_1, \dots, \lambda_N$, the following inequalities hold:*

$$(3.1.4) \qquad \sum_{k=1}^{\infty} k \left| \sum_{n=1}^{N} \beta_{nk} \lambda_n \right|^2 \leq \sum_{n=1}^{N} n |\lambda_n|^2,$$

or equivalently,

$$(3.1.5) \qquad \left| \sum_{n=1}^{N} \sum_{k=1}^{N} k \beta_{nk} \lambda_n \lambda_k \right| \leq \sum_{n=1}^{N} n |\lambda_n|^2.$$

Equalities hold in (3.1.4) and (3.1.5) if and only if $g \in \tilde{\Sigma}$.

Obviously (3.1.4) is equivalent to the Strong Grunsky Inequality (1.2.3), and (3.1.5) is equivalent to the Weak Grunsky Inequality (1.2.2).

The proof of the theorem requires a generalization of the Area Principle (cf. Theorem 1.1.2).

Lemma 3.1.1 (Generalized Area Principle). *Let $g \in \Sigma$, and p be an arbitrary non-constant polynomial of degree N,*

$$p(g(z)) = \sum_{k=-N}^{\infty} c_k z^{-k}, \qquad |z| > 1.$$

Then

$$(3.1.6) \qquad \sum_{k=-N}^{\infty} k |c_k|^2 \leq 0.$$

The equality holds if and only if $g \in \tilde{\Sigma}$.

Proof. Let C_r be the image of $|z| = r$ (> 1) under g, E_r the interior of C_r. Then by Green Theorem

$$0 \leq \iint_{E_r} |p'(w)|^2 \, du \, dv = \frac{1}{2\pi} \int_{C_r} \overline{p(w)} p'(w) \, dw$$

$$= \frac{1}{2\pi} \int_{|z|=r} \overline{p(g(z))} p'(g(z)) g'(z) \, dz$$

$$= -\pi \sum_{k=-N}^{\infty} k|c_k|^2 r^{-2k},$$

where $w = u + iv$. Let $r \to 1$, then

$$\sum_{k=-N}^{\infty} k|c_k|^2 = -\frac{2}{\pi} \iint_E |p'(w)|^2 \, du \, dv \leq 0.$$

The equality holds if and only if the measure of E is zero.

We now turn to prove Theorem 3.1.1.

Let $p(w) = \sum_{n=1}^{N} \lambda_n F_n(w)$. This is an n-th degree polynomial. From (3.1.3), we have

$$p(g(z)) = \sum_{n=1}^{N} \lambda_n z^n + \sum_{k=1}^{\infty} \sum_{n=1}^{N} \lambda_n \beta_{nk} z^{-k}.$$

By Lemma 3.1.1, we have (3.1.4).

Let $\nu_k = \sum_{n=1}^{N} \beta_{nk} \lambda_n$, $k = 1, \dots, N$. Then (3.1.4) can be written as

$$\sum_{k=1}^{\infty} k|\nu_k|^2 \leq \sum_{n=1}^{N} n|\lambda_n|^2.$$

By the Schwarz Inequality, we have

$$\left| \sum_{n=1}^{N} \sum_{k=1}^{N} k \beta_{nk} \lambda_n \mu_k \right|^2 = \left| \sum_{k=1}^{N} k \nu_k \mu_k \right|^2$$

$$\leq \sum_{k=1}^{N} k|\nu_k|^2 \sum_{k=1}^{N} k|\mu_k|^2 \leq \sum_{n=1}^{N} n|\lambda_n|^2 \sum_{k=1}^{N} k|\mu_k|^2.$$

This is the generalized Weak Grunsky Inequality (1.2.4).

If $\mu_k = \lambda_k$, the above inequality is reduced to (3.1.5). Thus the Strong Grunsky Inequality implies the generalized Weak Grunsky Inequality and the Weak Grunsky Inequality.

If $g \in \Sigma$, then we have the Strong Grunsky Inequality (3.1.4), and hence the generalized Weak Grunsky Inequality (3.1.5) and the Weak Grunsky Inequality (1.2.4).

Conversely, if $g(z) = z + b_0 + b_1 z^{-1} + b_2 z^{-2} + \cdots$ is holomorphic in $1 < |z| < \infty$, and if it has a pole at ∞ with residue 1, then for large enough $|z|$ and $|\zeta|$, $\log \dfrac{g(\zeta) - g(z)}{\zeta - z}$ is defined, and can be used to define γ_{nk} through (1.2.1). If such γ_{nk}'s satisfy the Grunsky Inequality (1.2.2) (equivalent to (3.1.5)), then we can prove that $g(z) \in \Sigma$.

Indeed, taking $\lambda_j = \delta_{nj}$ (Kronecker δ) in (1.2.2), then $|\gamma_{nn}| \leq \dfrac{1}{n}$. Taking $\lambda_j = \delta_{nj} + \delta_{kj}$, $k \neq n$, we find that

$$|\gamma_{nk}| \leq \frac{1}{n} + \frac{1}{k}.$$

Thus each γ_{nk} is bounded. And the series $\sum_{n=1}^{\infty} \sum_{k=1}^{\infty} \gamma_{nk} z^{-k} \zeta^{-n}$ is convergent in $|z| > 1$, $|\zeta| > 1$. Hence $\log \dfrac{g(\zeta) - g(z)}{\zeta - z}$ is holomorphic in $|z| > 1$, $|\zeta| > 1$. It implies that $g(z)$ is univalent in $|z| > 1$.

This shows that (3.1.5) implies the univalency of $g(z)$. Since the univalency of $g(z)$ implies the Strong Grunsky Inequality (3.1.4), we conclude that the Weak Grunsky Inequality implies the Strong Grunsky Inequality. Thus the three inequalities (1.2.2), (1.2.3) and (1.2.4) are equivalent.

The equality part of the theorem can be derived from the corresponding part of Lemma 3.1.1.

If we write $c_{nk} = -\sqrt{nk}\gamma_{nk}$, then (1.2.3) can be written as

$$\sum_{k=1}^{\infty} \left| \sum_{n=1}^{N} c_{kn} \lambda_n \right|^2 \leq \sum_{n=1}^{N} |\lambda_n|^2.$$

Let $\lambda = (\lambda_1, \ldots, \lambda_N)$, and $C_N^{\infty} = \begin{pmatrix} c_{11} & \cdots & c_{k1} & \cdots \\ \vdots & \ddots & \vdots & \ddots \\ c_{1N} & \cdots & c_{kN} & \cdots \end{pmatrix}$, an N row by ∞ column matrix, then the preceding inequality can be written as follows:

$$\lambda C_N^{\infty} \bar{C}_N^{\infty \prime} \bar{\lambda}' \leq \lambda \bar{\lambda}'.$$

Since λ is an arbitrary non-zero vector, the inequality is equivalent to

$$(3.1.7) \qquad\qquad I^{(N)} - C_N^\infty \bar{C}_N^{\infty\prime} \geq 0,$$

where $I^{(N)}$ is the N-th order identity matrix. This can be viewed as another form of the Strong Grunsky Inequality.

The equality in (3.1.7) holds if and only if $g \in \tilde{\Sigma}$. Hence C_N^∞ is unitary if and only if $g \in \tilde{\Sigma}$. In this case, $\sum_{n=1}^\infty c_{jn}\bar{c}_{nk} = \delta_{jk}$, $j, k = 1, 2, \ldots, N$, where N is an arbitrary positive integer. Thus $\{c_j = (c_{j1}, c_{j2}, \ldots)\}$, $j = 1, 2, \ldots$ forms an orthonormal system.

(1.2.3) implies

$$\sum_{k=1}^N \left| \sum_{n=1}^N c_{kn}\lambda_n \right|^2 \leq \sum_{n=1}^N |\lambda_n|^2.$$

Denote $C_N = \begin{pmatrix} c_{11} & \cdots & c_{N1} \\ \vdots & \ddots & \vdots \\ c_{1N} & \cdots & \cdots \end{pmatrix}$. The above inequality becomes

$$\lambda C_N \bar{C}_N' \lambda' \leq \lambda\lambda'.$$

Hence $I^{(N)} - C_N \bar{C}_N' \geq 0$. Since C_N is symmetric,

$$2I^{(N)} - 2\mathrm{Re}C_N = (I^{(N)} - C_N)(I^{(N)} - \bar{C}_N')$$
$$+ I^{(N)} - C_N \bar{C}_N' \geq 0.$$

Thus $I^{(N)} - \mathrm{Re}C_N \geq 0$. This inequality is the starting point of the proof of $|a_4| \leq 4$ and $|a_6| \leq 6$. The proof of $|a_4| \leq 4$ will be given in the next section.

From the relationships between c_{nk}, γ_{nk}, β_{nk}, b_j and a_j, it can be seen that a_j is a holomorphic function of c_{nk}. If we view c_{nk} as complex variables, then $I^{(N)} - C_N \bar{C}_N' \geq 0$ is a domain in $\mathbb{C}^{\frac{1}{2}N(N+1)}$. The modulus of holomorphic functions on this domain attain their maximum only at the characteristic manifold of the domain: $I^{(N)} - C_N \bar{C}_N' = 0$, i.e., all the symmetric unitary matrices. Hence, $g \in \tilde{\Sigma}$, and f is a slit mapping. The domain $I^{(N)} - C_N \bar{C}_N' > 0$ is in fact the classical domain of type II of several complex variables as defined by Hua. Interested readers are referred to the famous monograph of Hua [1] for detailed studies.

§3.2. The proof of $|a_4| \leq 4$ and related results

As mentioned in §1.2, $|a_4| \leq 4$ was proved by Garabedian and Schiffer [1] in 1955 using a variational method. However, their proof is long and complicated. In 1960, Charzynski and Schiffer gave a very simple proof of $|a_4| \leq 4$ using the Grunsky Inequality. In this section we will give their proof (Charzynski and Schiffer [1], Gong [2]). We will also give a proof of the Baranova Inequality $|a_4| \leq \frac{4}{15}(11+2|a_2|)$(V. A. Baranova [1]). The proof of $|a_6| \leq 6$ by Pederson and Ozawa follows the same line of argument. The proof of $|a_6| \leq 6$ is longer, and we omit the details of the proof. (cf. R.N. Pederson [1], M. Ozawa [1], S. Gong [2]).

Lemma 3.2.1. *If* $B = B^{(n)} = \begin{pmatrix} b_1 & v \\ v' & B_1 \end{pmatrix}$ *is a real symmetric positive semi-definite matrix, where* $b_1 = b_1^{(s)}$ *is an* $s \times s$ *real symmetric positive semi-definite matrix,* $v = v^{s \times (n-s)}$ *is an* $s \times (n-s)$ *matrix, and* $B_1 = B_1^{(n-s)}$ *is an* $(n-s) \times (n-s)$ *real symmetric positive semi-definite matrix, then* $B_1 - v'b_1^{-1}v$ *is a real symmetric positive semi-definite matrix.*

Proof. Since

$$\begin{pmatrix} I^{(s)} & 0 \\ -v'b_1^{-1} & I^{(n-s)} \end{pmatrix} \begin{pmatrix} b_1 & v \\ v' & B_1 \end{pmatrix} \begin{pmatrix} I^{(s)} & -b_1^{-1}v \\ 0 & I^{(n-s)} \end{pmatrix}$$

$$= \begin{pmatrix} b_1 & 0 \\ 0 & B_1 - v'b_1^{-1}v \end{pmatrix} \geq 0,$$

we have $B_1 - v'b_1^{-1}v \geq 0$.

We now turn to the proof that $|a_4| \leq 4$.

If $f(z) \in S$, then $g(\zeta) = \left(f\left(\frac{1}{\zeta}\right) \right)^{-1} \in \Sigma$, where $\zeta = \frac{1}{z}$. The function $F_2(\zeta) = \sqrt{g(\zeta^2)}$ is an odd function in Σ. For $|\zeta_\mu| > 1$, $|\zeta_\nu| > 1$, we have

$$\log \frac{F_2(\zeta_\mu) - F_2(\zeta_\nu)}{\zeta_\mu - \zeta_\nu} = -\sum_{m=1}^{\infty} \sum_{\ell=1}^{\infty} \gamma_{ml}\zeta_\mu^{-m}\zeta_\nu^{-\ell}.$$

Let $c_{k\ell} = -\sqrt{k\ell}\gamma_{k\ell}$, $\mathrm{Re} c_{k\ell} = p_{k\ell}$. Then from the last inequality from the previous section, $I^{(N)} - \mathrm{Re} C_N \geq 0$, we have

(3.2.1)
$$\begin{pmatrix} 1-p_{11} & -p_{13} & \cdots & -p_{1,2N+1} \\ -p_{13} & 1-p_{33} & \cdots & -p_{3,2N+1} \\ \vdots & \vdots & \ddots & \vdots \\ -p_{1,2N+1} & -p_{3,2N+1} & \cdots & 1-p_{2N+1,2N+1} \end{pmatrix} \geq 0.$$

For $N = 1$ the inequality becomes

$$\begin{pmatrix} 1 - p_{11} & -p_{13} \\ -p_{13} & 1 - p_{33} \end{pmatrix} \geq 0.$$

Using Lemma 3.2.1, we find that

$$p_{33} \leq 1 - \frac{p_{13}^2}{1 - p_{11}}.$$

We know that

$$p_{11} = \frac{1}{2}\mathrm{Re}a_2, \qquad p_{13} = \frac{\sqrt{3}}{2}\mathrm{Re}\left(a_3 - \frac{3a_2^2}{4}\right),$$

$$p_{33} = \mathrm{Re}\left(\frac{3}{2}a_4 - 3a_2a_3 + \frac{13}{8}a_2^3\right).$$

Hence the above inequality is equivalent to

$$\mathrm{Re}\left(\frac{3}{2}a_4 - 3a_2a_3 + \frac{13}{8}a_2^3\right) \leq 1 - \frac{\frac{3}{4}\left(\mathrm{Re}\left(a_3 - \frac{3a_2^2}{4}\right)\right)^2}{1 - \frac{1}{2}\mathrm{Re}a_2}.$$

Take $\lambda = a_3 - \frac{3}{4}a_2^2$, then

$$(3.2.2) \qquad \mathrm{Re}a_4 \leq \frac{2}{3} + \mathrm{Re}(2a_2\lambda) + \frac{5}{12}\mathrm{Re}a_2^3 - \frac{(\mathrm{Re}\lambda)^2}{2 - \mathrm{Re}a_2}.$$

Considering $e^{-i\alpha}f(e^{i\alpha}z)$ for an appropriate α, we may assume, without loss of generality, that $a_4 \geq 0$ and $\mathrm{Re}\{a_2\} \geq 0$.

Denote $\gamma_{11} = p + ix'$, $\gamma_{13} = \frac{1}{2}\lambda = y + iy'$. Then (3.2.2) becomes

$$a_4 \leq \frac{2}{3} + 8(yp - y'x' + \frac{10}{3}p(p^2 - 3x'^2) - \frac{2y^2}{1 - p}$$

$$= 4 - \frac{10}{3}(1 - p^3) + 8yp - 8y'x' - 10px'^2 - \frac{2y^2}{1 - p}$$

$$= 4 - \frac{10}{3}(1 - p^2) - \frac{10}{3}p^2(1 - p) + 8yp - 8y'x' - 10px'^2 - \frac{2y^2}{1 - p}.$$

By the Area Principle (Theorem 1.1.2), we have

$$1 - p^2 - x'^2 - 3y^2 - 3y'^2 \geq 0.$$

Therefore,

$$a_4 - 4 \leq \frac{-10}{3}(x'^2 + 3y'^2 + 3y^2) - \frac{10}{3}p^2(1-p)$$

$$+ 8yp - 8y'x' - 10px'^2 - \frac{2y^2}{1-p}$$

$$= -(x', y') \begin{pmatrix} \frac{10}{3} & +10p & 4 \\ 4 & 10 \end{pmatrix} \begin{pmatrix} x' \\ y' \end{pmatrix}$$

$$- (p, y) \begin{pmatrix} \frac{10}{3}(1-p) & -4 \\ -4 & \frac{2}{1-p} + 10 \end{pmatrix} \begin{pmatrix} p \\ y \end{pmatrix}.$$

Since the matrix

$$\begin{pmatrix} \frac{10}{3} + 10p & 4 \\ 4 & 10 \end{pmatrix}$$

is positive definite, and the matrix

$$\begin{pmatrix} \frac{10}{3}(1-p) & -4 \\ -4 & \frac{2}{1-p} + 10 \end{pmatrix}$$

is positive definite for $p < \frac{18}{25}$, we have $a_4 - 4 < 0$ when $p < \frac{18}{25}$.
    Since

$$8yp - \frac{2y^2}{1-p} \leq 8p^2(1-p),$$

hence we have

$$a_4 \leq 4 + 8p^2(1-p) - \frac{10}{3}(1-p^3) - 10px'^2 - 8y'x'$$

$$= 4 - \frac{10}{3} + 8p^2 - \frac{14}{3}p^3 - 10x'^2 - 8y'x'.$$

Applying the Area Principle again, we get, for $\alpha > 0$,

$$a_4 \leq 4 - \frac{10}{3} + 8p^2 - \frac{14}{3}p^3 + \alpha(1-p^2) - \alpha x'^2 - 3\alpha y'^2 - 10px'^2 - 8y'x'.$$

Thus

$$a_4 - 4 \leq \frac{-10}{3} + \alpha + (8 - \alpha)p^2 - \frac{14}{3}p^3 - (x', y') \begin{pmatrix} \alpha + 10p & 4 \\ 4 & 3\alpha \end{pmatrix} \begin{pmatrix} x' \\ y' \end{pmatrix}.$$

Taking $\alpha = \frac{5}{8}$, then for $p \geq \frac{18}{25}$, the matrix

$$\begin{pmatrix} \alpha + 10p & 4 \\ 4 & 3\alpha \end{pmatrix}$$

is positive definite, and moreover we have

$$-\frac{10}{3} + \alpha + (8 - \alpha)p^2 - \frac{14}{3}p^3 = -\frac{65}{24} + \frac{59}{8}p^2 - \frac{14}{3}p^3$$

$$= (1 - p)\left(-\frac{65}{24} - \frac{64}{24}p + \frac{14}{3}p^2\right).$$

Since $-\frac{65}{24} - \frac{64}{24}p + \frac{14}{3}p^2$ is an increasing function of p when $1 \geq p \geq \frac{18}{25}$, and

$$-\frac{65}{24} - \frac{64}{24} + \frac{14}{3} = -\frac{17}{24} < 0,$$

we have, for $1 \geq p \geq \frac{18}{25}$,

$$-\frac{65}{24} - \frac{59}{8}p^2 - \frac{14}{3}p^3 \leq 0.$$

Thus for $p \geq \frac{18}{25}$ we also have $a_4 \leq 4$. The equality holds if and only if $p = 1$, or equivalently, $a_2 = 2$, that is, if the function is the Koebe function.

Thus the proof of $|a_4| \leq 4$ is complete.

Moreover, we can use Grunsky inequality to get other deeper results. In 1970, Friedland used Grunsky inequality to prove the Robertson conjecture in the case of $n = 4$. In fact, in the process of the proof, Friedland proved the following result.

If $f(z) = z + \sum_{n=2}^{\infty} a_n z_n \in S$, then

(3.2.3) $$|a_4| \leq \frac{7}{2} + \frac{1}{8}|a_2|^2 \leq 4.$$

Thus $|a_4| = 4$ if and only if $|a_2| = 2$, that is, $f(z)$ is the Koebe function or its rotation. Hence the equality of the Robertson conjecture in the case $n = 4$ holds if and only if $f_2(z) = \frac{z}{1-z^2}$ or its rotation.

We can get more precise formulas than (3.2.3). For example, in 1971, Baranova (V.A.Baranova [1]) proved the following formula: $|a_4| \leq \frac{4}{15}(11 +$

$2|a_2|$). The proof uses the Grunsky inequalities for function in S which are now described.

If $f(z) \in S$, let

$$\log \frac{f(\xi) - f(z)}{\xi - z} = \sum_{p=0}^{\infty} \sum_{q=0}^{\infty} \omega_{p,q} \xi^p z^q,$$

This formula corresponds to (1.2.1) in the case $g(\zeta) \in \Sigma$. Then we can prove the corresponding formula of (1.2.2),

$$(3.2.4) \qquad \left| \sum_{p=1}^{\infty} \sum_{q=1}^{\infty} \omega_{p,q} x_p x_q \right| \leq \sum_{p=1}^{\infty} \frac{1}{p} |x_p|^2,$$

where $x_1, x_2, \cdots$ are arbitrary complex numbers, such that the right hand side of (3.2.4) converges.

The formula corresponding to (1.2.3) is

$$(3.2.5) \qquad \sum_{p=1}^{\infty} p \left| \sum_{q=1}^{\infty} \omega_{p,q} x_q \right|^2 \leq \sum_{p=1}^{\infty} \frac{1}{p} |x_p|^2,$$

where $x_1, x_2, \cdots$ are arbitrary complex numbers, such that the right hand side of (3.2.5) converges.

The proofs of (3.2.4) and (3.2.5) are similar to the proofs of (1.2.2) and (3.1.4). As for functions in Σ, (3.2.4) and (3.2.5) are called the Grunsky inequalities again. And it can be shown that they are equivalent.

We now prove Baranova's bound on $|a_4|$.

Just as in the Σ case, we consider the function $\sqrt{f(z^2)} = h(z)$ for (3.2.5). First letting $x_1 = 1$, $x_2 = x_3 = \cdots = 0$, and then letting $x_1 = \ell$, $x_2 = 0$, $x_3 = 2$, $x_4 = x_5 = \cdots = 0$, we have

$$(3.2.6) \qquad |\omega_{11}|^2 + 3|\omega_{13}|^2 \leq 1,$$

$$(3.2.7) \qquad |\omega_{11}\ell + 2\omega_{13}|^2 + 3|\omega_{31}\ell + 2\omega_{33}|^2 \leq |\ell|^2 + \frac{4}{3}.$$

Direct computation yields

$$\omega_{11} = \frac{1}{2}a_2, \quad \omega_{13} = \omega_{31} = \frac{1}{2}\left(a_3 - \frac{3}{4}a_2^2\right),$$

$$\omega_{33} = \frac{1}{2}a_4 - a_2\left(a_3 - \frac{13}{24}a_2^2\right) = \frac{1}{2}a_4 - 4\omega_{11}\omega_{13} - \frac{5}{3}\omega_{11}^3.$$

Substitute the above into (3.2.7), we get

$$3\left|a_4 - (8\omega_{11} - \ell)\omega_{13} - \frac{10}{3}\omega_{11}^3\right|^2 + |\omega_{11}\ell + 2\omega_{13}|^2$$
$$\leq |\ell|^2 + \frac{4}{3}.$$

Applying a rotation if necessary, we may assume that $a_4 > 0$. The above inequality immediately implies

$$|a_4| \leq \left(\frac{4}{9} + \frac{1}{3}|\ell|^2 - \frac{1}{3}|\omega_{11}\ell + 2\omega_{13}|^2\right)^{\frac{1}{2}}$$
$$+ \operatorname{Re}\left\{(8\omega_{11} - \ell)\omega_{13} + \frac{10}{3}\omega_{11}^3\right\}.$$

The right hand side of the above inequality is less than or equal to

$$\frac{2}{3} + \frac{1}{4}|\ell|^2 - \frac{1}{4}|\omega_{11}\ell + 2\omega_{13}|^2 + \operatorname{Re}\left\{(8\omega_{11} - \ell)\omega_{13} + \frac{10}{3}\omega_{11}^3\right\}$$
$$= \frac{2}{3} + \frac{1}{4}(1 - |\omega_{11}|^2)|\ell|^2 - |\omega_{13}|^2 + \operatorname{Re}\left\{(8\omega_{11} - \ell - \bar{\omega}_{11}\bar{\ell})\omega_{13} + \frac{10}{3}\omega_{11}^3\right\}.$$

Write

$$\omega_{11} = xe^{i\phi}, \quad x > 0, \quad -\pi < \phi \leq \pi,$$
$$\ell = \frac{8x}{1+x}e^{-\frac{1}{2}\phi i}\cos\frac{3}{2}\phi, \quad y = \left|\sin\frac{3}{2}\phi\right|,$$

then the above right hand side becomes

$$\frac{2}{3} + \frac{1}{4}(1-x^2)\frac{64x^2}{(1+x)^2}\cos^2\frac{3}{2}\phi - |\omega_{13}|^2 + \mathrm{Re}\left\{8xe^{i\phi}\omega_{13}\right.$$

$$-\frac{8x}{1+x}e^{-\frac{1}{2}\phi i}\cos\frac{3}{2}\phi\omega_{13} - xe^{-i\phi}\frac{8x}{1+x}e^{\frac{1}{2}\phi i}\cos\frac{3}{2}\phi\omega_{13}$$

$$\left. +\frac{10}{3}x^3 e^{3\phi i}\right\}$$

$$=\frac{3}{2} + \frac{16x^2(1-x)}{1+x}\left(1 - \sin^2\frac{3}{2}\phi\right)$$

$$-|\omega_{13}|^2 + 8x\mathrm{Re}\{e^{i\phi}\omega_{13}\} - \frac{8x}{1+x}\cos\frac{3}{2}\phi\mathrm{Re}\{e^{-\frac{1}{2}\phi i}\omega_{13}\}$$

$$-\frac{8x^2}{1+x}\cos\frac{3}{2}\phi\mathrm{Re}\{e^{-\frac{1}{2}\phi i}\omega_{13}\} + \frac{10}{3}x^3\cos 3\phi$$

$$=\frac{2}{3} + 16x^2\frac{1-x}{1+x} - |\omega_{13}|^2 - \frac{16x^2(1-x)}{1+x}y^2$$

$$+ 8x\mathrm{Re}\left\{e^{-\frac{1}{2}\phi i}i\sin\frac{3}{2}\phi\omega_{13}\right\} + \frac{10}{3}x^3 - \frac{20}{3}x^3 y^2$$

$$\le \frac{2}{3} + 16x^2\frac{1-x}{1+x} + \frac{10}{3}x^3 - |\omega_{13}|^2 + 8x|\omega_{13}|y$$

$$-\left(\frac{20}{3}x^3 + 16x^2\frac{1-x}{1+x}\right)y^2.$$

Taking the supremum of the above expression in y, and using (3.2.6), $|\omega_{13}| \le \left(\frac{1-x^2}{3}\right)^{\frac{1}{2}}$, we arrive at the inequality

$$|a_4| \le \frac{2}{3} + 16x^2\frac{1-x}{1+x} + \frac{10}{3}x^3$$

(3.2.8)
$$+ \frac{1}{3}\frac{19 + 14x - 5x^2}{12 - 7x + 5x^2}(1-x),$$

where $x = \frac{1}{2}|a_2|$.

Since $\dfrac{2x}{1+x} \le \dfrac{1+x}{2}$, and for $0 \le x \le 1$

$$\frac{19 + 14x - 5x^2}{12 - 7x + 5x^2} \le 3.24 - 0.44x,$$

it follows

(3.2.9)
$$|a_4| \le \frac{1}{3}(2 + 15.24x - 3.68x^2 - 1.56x^3).$$

Addition of $\frac{1.56}{3}x(1-x)^2$ to this inequality yields

$$(3.2.10) \qquad\qquad |a_4| \le \frac{1}{3}(2 + 16.8x - 6.8x^2).$$

Again addition of $\frac{6.8}{3}(1-x)^2$ yields

$$(3.2.11) \qquad\qquad |a_4| \le \frac{4}{15}(11 + 4x).$$

The equality holds in (3.2.8), (3.2.9), and (3.2.10) when $a_2 = 0$ or $|a_2| = 2$, with functions $f(z) = \dfrac{z}{(1-\eta z^3)^{3/2}}$ and $f(z) = \dfrac{z}{(1-\eta z)^2}$, $|\eta| = 1$.

§3.3. Lebedev-Milin Inequalities

In this section we will prove the three very important Lebedev-Milin Inequalities which mentioned in §1.3. Then we use these three inequalities to "exponentiate" the Grunsky inequality. Based upon these results, we will give the proofs of Milin Theorem, Milin Lemma, Bazilevich Inequality and Hayman Regularity Theorem as we mentioned at §1.2 and §1.3. Assume $\phi(z) = \sum_{k=1}^{\infty} \alpha_k z^k$ is an arbitrary power series with a positive radius of convergence and satisfies $\phi(0) = 0$. Write

$$(1.3.4) \qquad\qquad e^{\phi(z)} = \sum_{k=0}^{\infty} \beta_k z^k, \qquad \beta_0 = 1.$$

Then we will prove that (1.3.5), (1.3.6) and (1.3.7) hold. We start with

First Lebedev-Milin Inequality. *If* $\sum_{k=1}^{\infty} k|\alpha_k|^2 < \infty$, *then the inequality*

$$(1.3.5) \qquad\qquad \sum_{k=0}^{\infty} |\beta_k|^2 \le \exp\left\{\sum_{k=1}^{\infty} k|\alpha_k|^2\right\},$$

holds, where the equality holds if and only if $\alpha_k = \gamma^k/k$, $k = 1, 2, \ldots$, γ *is a complex number with* $|\gamma| < 1$.
Proof. Differentiating $\psi(z) = e^{\phi(z)}$, we get $\psi'(z) = \psi(z)\phi'(z)$. Hence the coefficients satisfy

$$(3.3.1) \qquad\qquad \beta_n = \frac{1}{n}\sum_{k=0}^{n-1}(n-k)\alpha_{n-k}\beta_k, \qquad \beta_0 = 1.$$

By Cauchy inequality

(3.3.2) $$|\beta_n|^2 \le \frac{1}{n} \sum_{k=0}^{n-1} (n-k)^2 |\alpha_{n-k}|^2 |\beta_k|^2.$$

Denote $a_k = k|\alpha_k|^2$ and define b_k recursively by

(3.3.3) $$b_n = \frac{1}{n} \sum_{k=0}^{n-1} (n-k) a_{n-k} b_k, \qquad b_0 = 1.$$

Then the inequality $|\beta_n|^2 \le b_n$, $n = 0, 1, 2, \ldots$ can be proved by induction. When $n = 0$, $b_0 = 1$ and $\beta_0 = 1$, the inequality holds trivially. Assume the inequality holds for integers $n \le m$, then by (3.3.2)

$$|\beta_{m+1}|^2 \le \frac{1}{m+1} \sum_{k=0}^{m} (m+1-k)^2 |\alpha_{m+1-k}|^2 |\beta_k|^2$$

$$\le \frac{1}{m+1} \sum_{k=0}^{m} (m+1-k)^2 |\alpha_{m+1-k}|^2 b_k$$

$$= \frac{1}{m+1} \sum_{k=0}^{m} (m+1-k) a_{m+1-k} b_k = b_{m+1}.$$

Comparing (3.3.3) and (3.3.1), we have

$$\sum_{k=0}^{\infty} b_k z^k = \exp \left\{ \sum_{k=1}^{\infty} a_k z^k \right\}.$$

Since $a_k \ge 0$ and $b_k \ge 0$, we have the inequality

$$\sum_{k=0}^{\infty} |\beta_k|^2 \le \sum_{k=0}^{\infty} b_k = \exp \left\{ \sum_{k=1}^{\infty} a_k \right\}$$

$$= \exp \left\{ \sum_{k=1}^{\infty} k |\alpha_k|^2 \right\}.$$

This proves (1.3.5). The equality holds if and only if $|\beta_k|^2 = b_k$ for all k, i.e., the equality of (3.3.2) holds for all n. The conditions for equality in the Cauchy inequality are

$$(n-k)\alpha_{n-k}\beta_k = \lambda_n, \qquad k = 0, 1, \cdots, n-1.$$

where λ_n, $n = 1, 2, \cdots$ are complex constants. Thus $n\alpha_n = \lambda_n$ since $\beta_0 = 1$. Substituting it into (3.3.1), we have $\beta_n = \lambda_n$. It implies $\lambda_{n-k}\lambda_k = \lambda_n$. It follows that $\lambda_2 = \lambda_1^2, \cdots, \lambda_n = \lambda_1^n$. Let $\lambda_1 = \gamma$, then $\alpha_n = \gamma^n/n$, $\beta_n = \gamma^n$. In other words, the equality holds if and only if

$$\varphi(z) = -\log(1 - \gamma z), \qquad \psi(z) = (1 - \gamma z)^{-1}.$$

To guarantee the convergence of $\sum k|\alpha_k|^2 < \infty$, the condition $|\gamma| < 1$ is needed.

Now we can prove:

Second Lebedev-Milin Inequality.   *For all positive integers $n = 1, 2, \cdots$, the inequality*

$$(1.3.6) \qquad \frac{1}{n+1} \sum_{k=0}^{n} |\beta_k|^2 \leq \exp\left\{ \frac{1}{n+1} \sum_{m=1}^{n} \sum_{k=1}^{m} \left(k|\alpha_k|^2 - \frac{1}{k} \right) \right\}$$

holds. For any fixed integer n, the equality holds if and only if $\alpha_k = \gamma^k/k$, $k = 1, 2, \cdots, n$, where γ is a complex constant with $|\gamma| = 1$.
Proof. Using Cauchy inequality on (3.3.1), we get

$$(3.3.4) \qquad n^2|\beta_n|^2 \leq \sum_{k=1}^{n} k^2|\alpha_k|^2 \sum_{k=0}^{n-1} |\beta_k|^2.$$

Let

$$A_n = \sum_{k=1}^{n} k^2|\alpha_k|^2, \qquad B_n = \sum_{k=0}^{n} |\beta_k|^2,$$

then

$$B_n = B_{n-1} + |\beta_n|^2 \leq \left\{ 1 + \frac{1}{n^2} A_n \right\} B_{n-1}$$

$$= \frac{n+1}{n} \left\{ 1 + \frac{A_n - n}{n(n+1)} \right\} B_{n-1}$$

$$\leq \frac{n+1}{n} B_{n-1} \exp\left\{ \frac{A_n - n}{n(n+1)} \right\}$$

by (3.3.4).

Inserting the formulas of $B_{n-1}, B_{n-2}, \cdots$ into the previous formula, we have

$$B_n \leq (n+1) \exp \left\{ \sum_{k=1}^{n} \frac{A_k - k}{k(k+1)} \right\}$$

$$= (n+1) \exp \left\{ \sum_{k=1}^{n} \frac{A_k}{k(k+1)} + 1 - \sum_{k=1}^{n+1} \frac{1}{k} \right\}.$$

Let $s_n = \sum_{k=1}^{n} \frac{1}{k(k+1)}$, then

$$\sum_{k=1}^{n} A_k \frac{1}{k(k+1)} = \sum_{k=1}^{n} A_k(s_k - s_{k-1})$$

$$= A_n s_n - A_n s_{n-1} + A_{n-1} s_{n-1} - \cdots$$

$$= A_n s_n - \sum_{k=1}^{n} (A_k - A_{k-1}) s_{k-1}$$

$$= A_n s_n - \sum_{k=1}^{n} k^2 |\alpha_k|^2 s_{k-1}$$

$$= \sum_{k=1}^{n} k |\alpha_k|^2 - \frac{1}{n+1} \sum_{k=1}^{n} k^2 |\alpha_k|^2$$

$$- \sum_{k=1}^{n} \frac{k(n+1-k)}{n+1} |\alpha_k|^2$$

since

$$s_n = \sum_{k=1}^{n} \frac{1}{k(k+1)} = \sum_{k=1}^{n} \left(\frac{1}{k} - \frac{1}{k+1} \right) = 1 - \frac{1}{n+1}.$$

We have

$$B_n \leq (n+1) \exp \left\{ \frac{1}{n+1} \sum_{k=1}^{n} (n+1-k) \left(k|\alpha_k|^2 - \frac{1}{k} \right) \right\}$$

(3.3.4')

$$= (n+1) \exp \left\{ \frac{1}{n+1} \sum_{m=1}^{n} \sum_{k=1}^{m} \left(k|\alpha_k|^2 - \frac{1}{k} \right) \right\}.$$

This proves (1.3.6). The equality in (1.3.6) holds if and only if the equality in the Cauchy inequality holds and the equality in $1 + x \leq e^x$ holds. Then

$A_k = k$ for all of $k = 1, 2, \cdots, n$, and there exist complex constants λ_m, $m = 0, 1, \cdots, n$, such that

$$\beta_k = \lambda_m(m - k)\bar{\alpha}_{m-k}, \qquad k = 0, 1, \cdots, m - 1.$$

Inserting it into (3.3.1), we find

$$m\beta_m = \lambda_m A_m = m\lambda_m,$$

that is, $\beta_m = \lambda_m$, $m = 1, 2, \cdots, n$. Since $\beta_0 = 1$, we have

$$\lambda_m m\bar{\alpha}_m = 1, \qquad m = 1, 2, \cdots, n$$

and

$$\lambda_1 = \beta_1 = \lambda_m(m - 1)\bar{\alpha}_{m-1} = \lambda_m/\lambda_{m-1}.$$

Then $\beta_m = \lambda_m = \lambda_1^m$ and $m\bar{\alpha}_m = \lambda_1^{-m}$, $m = 1, \cdots, n$. Finally, $|\lambda_1| = 1$ is true since $A_k = k$ holds for all k.

Finally, we prove

Third Lebedev-Milin Inequality. *For all integers $n = 1, 2, \cdots$, the inequality*

(1.3.7)
$$|\beta_n|^2 \le \exp\left\{\sum_{k=1}^{n}\left(k|\alpha_k|^2 - \frac{1}{k}\right)\right\},$$

holds. The equality holds if and only if $\alpha_k = \frac{\gamma^k}{k}$, $k = 1, 2, \cdots, n$ where γ is a complex number with $|\gamma| = 1$.
Proof. Since (3.3.4) and (1.3.6), we get

$$
\begin{aligned}
|\beta_n|^2 &\le \frac{1}{n}\sum_{k=1}^{n} k^2|\alpha_k|^2 \exp\left\{\frac{1}{n}\sum_{m=1}^{n-1}\sum_{k=1}^{m}\left(k|\alpha_k|^2 - \frac{1}{k}\right)\right\} \\
&= \frac{1}{n}\sum_{k=1}^{n} k^2|\alpha_k|^2 \exp\left\{\frac{1}{n}\sum_{k=1}^{n-1}(n - k)\left(k|\alpha_k|^2 - \frac{1}{k}\right)\right\} \\
&= e\frac{A_n}{n}\exp\left\{-\frac{A_n}{n} + \sum_{k=1}^{n-1}\left(k|\alpha_k|^2 - \frac{1}{k}\right)\right\}.
\end{aligned}
$$

The inequality $xe^{-x} \le \frac{1}{e}$ holds for all x. Taking $x = \frac{A_n}{n}$, we obtain (1.3.7). The equality holds when the equality in (3.3.4) holds for n, and the equality in (1.3.6) holds for $n-1$. It reduces to $\alpha_k = \frac{\gamma^k}{k}$, $k = 1, 2, \cdots, n$, where γ is a complex number with $|\gamma| = 1$.

Starting from these inequalities, we can "exponentiate" the Grunsky Inequality. Let $g \in \Sigma$, and

$$\log \frac{g(z) - g(\zeta)}{z - \zeta} = -\sum_{n=1}^{\infty} \sum_{k=1}^{\infty} \gamma_{nk} \zeta^{-k} z^{-n}$$

(3.3.5)
$$= -\sum_{n=1}^{\infty} A_n \left(\frac{1}{\zeta} \right) z^{-n},$$

where $A_n(w) = \sum_{k=1}^{\infty} \gamma_{nk} w_k$, $|w| < 1$. Taking $\lambda_k = w^k$ in the strong Grunsky Inequality

(1.2.3)
$$\sum_{n=1}^{\infty} n \left| \sum_{k=1}^{\infty} \gamma_{nk} \lambda_k \right|^2 \le \sum_{n=1}^{\infty} \frac{1}{n} |\lambda_n|^2,$$

then (1.2.3) becomes

(3.3.6)
$$\sum_{n=1}^{\infty} n |A_n(w)|^2 \le \sum_{n=1}^{\infty} \frac{1}{n} |w|^{2n} = -\log(1 - |w|^2).$$

The ν-th derivative of $A(w)$ is

$$A^{(\nu)}(w) = \sum_{k=\nu}^{\infty} \gamma_{nk} k(k-1) \cdots (k-\nu+1) w^{k-\nu}, \qquad \nu = 1, 2, \cdots.$$

Taking

$$\lambda_k = \begin{cases} 0, & k < \nu; \\ k(k-1) \cdots (k-\nu+1) w^{k-\nu}, & k \ge \nu; \end{cases}$$

we have

$$\sum_{n=1}^{\infty} n |A_n^{(\nu)}(w)|^2 \le \sum_{k=\nu}^{\infty} \frac{1}{k} k^2 (k-1)^2 \cdots (k-\nu+1)^2 |w|^{2(k-\nu)}$$

$$= - \left[\frac{\partial^{2\nu}}{\partial w^\nu \partial \bar{z}^\nu} \log(1 - w\bar{z}) \right]_{z=w}.$$

When $\nu = 1$, it is

$$\sum_{n=1}^{\infty} n|A_n'(w)|^2 \leq (1 - |w|^2)^{-2}.$$

Let

$$(3.3.7) \qquad \exp\left\{\sum_{n=1}^{\infty} A_n(w)z^n\right\} = \sum_{n=0}^{\infty} B_n(w)z^n.$$

Using the first Lebedev-Milin inequality, we have

$$\sum_{n=0}^{\infty} |B_n(w)|^2 \leq \exp\left\{\sum_{n=1}^{\infty} n|A_n(w)|^2\right\}.$$

By (3.3.6), we obtain

$$\sum_{n=0}^{\infty} |B_n(w)|^2 \leq (1 - |w|^2)^{-1}, \quad |w| < 1.$$

We know

$$\frac{g'(z)}{g(z) - w} = \sum_{n=0}^{\infty} F_n(w)z^{-n-1},$$

by (3.1.1), where $F_n(w)$ is the n-th Faber polynomial. Integrating the previous equality with respect to z, and taking the integrating constant such that it is zero when $z = \infty$, we have

$$(3.3.8) \qquad \log\frac{z}{g(z) - w} = \sum_{n=1}^{\infty} \frac{1}{n}F_n(w)z^{-n}.$$

From (3.3.5), we know

$$\log\frac{z}{g(z) - g(\zeta)} = \sum_{n=1}^{\infty} A_n\left(\frac{1}{\zeta}\right)z^{-n} + \log\frac{z}{z - \zeta}.$$

Comparing these two formulas, we have

$$(3.3.9) \qquad F_n(g(\zeta)) = nA_n\left(\frac{1}{\zeta}\right) + \zeta^n, \quad n = 1, 2, \cdots.$$

These state the relationship between the Faber polynomials and $A_n's$. The relationship between the Faber polynomials and $B_n's$ can be derived as follows:

From (3.3.5) and (3.3.7), we have

$$\frac{z - \zeta}{g(z) - g(\zeta)} = \sum_{n=0}^{\infty} B_n \left(\frac{1}{\zeta}\right) z^{-n}, \quad |z| > 1, |\zeta| > 1.$$

Differentiating (3.3.8) with respect to w, we note

$$\frac{1}{g(z) - w} = \sum_{n=1}^{\infty} \frac{1}{n} F_n'(w) z^{-n}.$$

Comparing these two formulas, we have

$$\sum_{n=1}^{\infty} \frac{1}{n} F_n'(g(\zeta)) z^{-n} = \sum_{k=0}^{\infty} \zeta^k z^{-k-1} \sum_{n=0}^{\infty} B_n \left(\frac{1}{\zeta}\right) z^{-n}.$$

Comparing the coefficients of the terms of z, we obtain

$$(3.3.10) \qquad F_n'(g(\zeta)) = n \sum_{k=0}^{n-1} \zeta^{n-k-1} B_k \left(\frac{1}{\zeta}\right).$$

Using the above results, we can prove the Milin Theorem in section 1.2: If $f \in S$, then $|a_n| < 1.243n$, $n = 2, 3, \cdots$. As we pointed out in the section 1.2, this is the first result about the estimate of the coefficients without using the estimation of the mean of modulus of function, that is, it is the first result of the estimate of coefficients that is better than $\frac{e}{2}n$.

Using Cauchy inequality on (3.3.10), we obtain

$$|F_n'(g(\zeta))|^2 \leq n^2 \sum_{k=0}^{n-1} |\zeta|^{2k} \sum_{k=0}^{n-1} |B_k \left(\frac{1}{\zeta}\right)|^2$$

$$\leq n^2 \frac{|\zeta|^{2n} - 1}{|\zeta|^2 - 1} \frac{|\zeta|^2}{|\zeta|^2 - 1}, \quad |\zeta| > 1.$$

If g maps $|\zeta| = \rho (> 1)$ to a Jordan curve C_ρ, which contains the origin, then

$$|F_n'(0)| \leq \max_{w \in C_\rho} |F_n'(w)| = \max_{|\zeta| = \rho} |F_n'(g(\zeta))|,$$

by the maximum modulus theorem. But $a_n = \frac{1}{n} F_n'(0)$, $n = 2, 3, \cdots$, and it follows that

$$|a_n| \leq \frac{\rho}{\rho^2 - 1} (\rho^{2n} - 1)^{\frac{1}{2}}, \quad \rho > 1.$$

Let $\rho = e^{\frac{x}{2n}}$. The previous inequality becomes

$$|a_n| \leq \frac{e^{\frac{x}{2n}}}{e^{\frac{x}{n}} - 1}(e^x - 1)^{\frac{1}{2}} = \frac{1}{e^{\frac{x}{2n}} - e^{-\frac{x}{2n}}}(e^x - 1)^{\frac{1}{2}}.$$

When $t > 0$,

$$\frac{e^t - e^{-t}}{2t} = 1 + \frac{1}{3!}t^2 + \cdots > 1.$$

Thus

$$|a_n| \leq \frac{(e^x - 1)^{\frac{1}{2}}}{x}n.$$

The function $x^{-1}(e^x - 1)^{\frac{1}{2}}$ takes minimum value at $x = \log 2 - \log(2 - x)$, where the value of x is approximately equal to 1.594, and substituting this value in the previous inequality, we conclude that

$$|a_n| < 1.243n, \quad n = 2, 3, \cdots.$$

Thus we have the following theorem(Milin [1,2]).
Milin Theorem.   *Let $f(z) \in S$, with expansion (1.1.5) then*

$$|a_n| < 1.243n, n = 2, 3, \cdots.$$

Now we prove the important Milin Lemma in §1.3.
Let $f \in S$, and let $\gamma_k's$ be the coefficients in

$$\log \frac{f(z)}{z} = 2\sum_{n=1}^{\infty} \gamma_n z^n, \quad |z| < 1.$$

From (3.3.8), we know $2\gamma_n = \frac{1}{n}F_n(0)$, then

$$4\sum_{k=1}^{n} k|\gamma_k|^2 = \sum_{k=1}^{n} \frac{1}{k}|F_k(0)|^2.$$

Using the inequality $(a + b)^2 \leq 2(a^2 + b^2)$, and (3.3.9), we have

$$\frac{1}{k}|F_k(g(\zeta))|^2 \leq 2k\left|A_k\left(\frac{1}{\zeta}\right)\right|^2 + \frac{2}{k}|\zeta|^{2n}.$$

By (3.3.6),

$$\sum_{k=1}^{n} \frac{1}{k}|F_k(g(\zeta))|^2 \leq 2\sum_{k=1}^{n} k\left|A_k(\frac{1}{\zeta})\right|^2 + 2\sum_{k=1}^{n} \frac{1}{k}|\zeta|^{2k}$$

$$\leq -2\log(1 - \frac{1}{|\zeta|^2}) + 2\sum_{k=1}^{n} \frac{1}{k}|\zeta|^{2k}.$$

As usual, we let $g(\zeta) = \left[f\left(\frac{1}{\zeta}\right)\right]^{-1}$ for $|\zeta| > 1$, then g is in Σ. Note that g maps $|\zeta| = \rho (> 1)$ to a Jordan curve C_ρ which contains the origin. The function $\sum_{k=1}^{n}|F_k(w)|^2$ is subharmonic in the domain bounded by C_ρ (cf. Ahlfors[1]), then

$$4\sum_{k=1}^{n} k|\gamma_k|^2 = \sum_{k=1}^{n} \frac{1}{k}|F_k(0)|^2 \leq \max_{w \in C_\rho} \sum_{k=1}^{n} \frac{1}{k}|F_k(w)|^2$$

(3.3.11)
$$\leq -2\log(1 - \rho^{-2}) + 2\sum_{k=1}^{n} \frac{\rho^{2k}}{k},$$

by the maximum modulus principle of subharmonic functions. Dividing by 2, we have

(3.3.11')
$$2\sum_{k=1}^{n} k|\gamma_k|^2 \leq \sum_{k=1}^{n} \frac{1}{k}\rho^{2k} - \log(1 - \rho^{-2}), \quad \rho > 1.$$

Taking $\rho^2 = e^t$, $t = \frac{2x}{2n+1}$, $t > 0$, we find

$$-\log(1 - \rho^{-2}) = \log(1 - e^{-t}) = \frac{t}{2} - \log(e^{\frac{t}{2}} - e^{-\frac{t}{2}}) < \frac{t}{2} - \log t$$

$$= \frac{x}{2n+1} - \log x + \log\left(n + \frac{1}{2}\right)$$

$$< \frac{x}{2n+1} - \log x + \sum_{k=1}^{n} \frac{1}{k} - \gamma,$$

where we use the inequality

(3.3.12)
$$\log\left(n + \frac{1}{2}\right) < \sum_{k=1}^{n} \frac{1}{k} - \gamma,$$

and γ is the Euler constant, $\gamma = 0.577\cdots$.

Obviously,

$$\sum_{k=1}^{n} \frac{1}{k}\rho^{2k} = \sum_{k=1}^{n} \frac{1}{k} \sum_{m=0}^{\infty} \frac{1}{m!}(kt)^m = \sum_{m=0}^{\infty} \frac{1}{m!} t^m \sum_{k=1}^{n} k^{m-1}.$$

Since

$$(3.3.13) \qquad m \sum_{k=1}^{n} k^{m-1} < \left(n + \frac{1}{2} \right)^m, \quad m = 1, 2, \cdots,$$

we have

$$\sum_{k=1}^{n} \frac{1}{k} \rho^{2k} < \sum_{k=1}^{n} \frac{1}{k} + nt + \sum_{m=2}^{\infty} \frac{1}{m!m} t^m (n + \frac{1}{2})^m$$

$$= \sum_{k=1}^{n} \frac{1}{k} + \frac{2nx}{2n+1} + \sum_{m=2}^{\infty} \frac{x^m}{m!m}.$$

Substituting all these results into (3.3.11'), we find that

$$2 \sum_{k=1}^{n} k|\gamma_k|^2 \leq \int_0^x \frac{e^y - 1}{y} dy - \log x + 2 \sum_{k=1}^{n} \frac{1}{k} - \gamma \equiv G_n(x),$$

To find the minimum value of G_n, we differentiate G_n, $G_n'(x) = \frac{e^x - 1}{x} - \frac{1}{x} = 0$. G_n takes its minimum value at $x = \log 2$. Thus

$$\sum_{k=1}^{n} k|\gamma_k|^2 \leq \frac{1}{2} G_n(\log 2) = \sum_{k=1}^{n} \frac{1}{k} + \delta.$$

where

$$\delta = \frac{1}{2} \int_0^{\log 2} \frac{e^y - 1}{y} dy - \frac{1}{2} \log \log 2 - \frac{\gamma}{2} < 0.312.$$

This proves Milin Lemma.

The proofs of (3.3.12) and (3.3.13) are as follows:

We prove (3.3.12) first. Let

$$y_n = \sum_{k=1}^{n} \frac{1}{k} - \log \left(n + \frac{1}{2} \right),$$

then $y_n - y_{n-1} = \psi(n)$, and

$$\psi(x) = \frac{1}{x} - \log \frac{x + \frac{1}{2}}{x - \frac{1}{2}}.$$

Obviously, $\psi(1) < 0$, $\psi'(x) > 0$, and $\psi(x) \to 0$ when $x \to +\infty$. Thus $y_n < y_{n-1}$, and y_n decrease to γ.

We prove (3.3.13) by induction.

Obviously, (3.3.13) is true when $n = 1$. If (3.3.13) is true for $n - 1$, then

$$\sum_{k=1}^{n} k^{m-1} < \frac{1}{m}\left(n - \frac{1}{2}\right)^m + n^{m-1} < \frac{1}{m}\left(n + \frac{1}{2}\right)^m,$$

since $(n + \frac{1}{2})^m - (n - \frac{1}{2})^m > mn^{m-1}$.

The Milin lemma motivated the very important Milin Conjecture (1.3.13).
Now we give the proof of the Bazilevich Inequality (1.3.9) in §1.3.

Let $f(z) \in S$, $g(\zeta) = \frac{1}{f(\frac{1}{\zeta})}$, then

$$\sum_{n=1}^{\infty} n\left|A_n(z) - \frac{1}{n}\bar{z}^n\right|^2 = \sum_{n=1}^{\infty} n|A_n(z)|^2 - 2Re\left\{\sum_{n=1}^{\infty} A_n(z)z^n\right\} + \sum_{n=1}^{\infty} \frac{1}{n}|z|^{2n}$$

(3.3.14)
$$\leq -2\log(1 - |z|^2) - 2Re\left\{\sum_{n=1}^{\infty} A_n(z)z^n\right\}$$

by (3.3.5) and (3.3.6). Using (3.3.14), the equality

$$-\sum_{n=1}^{\infty} A_n(z)z^n = \log g'(\zeta) = \log\frac{z^2 f'(z)}{(f(z))^2},$$

and inequality $\left|\frac{zf'(z)}{f(z)}\right| \leq \frac{1+r}{1-r}$ hold, we have

$$\sum_{n=1}^{\infty} n\left|A_n(z) - \frac{1}{n}\bar{z}^n\right|^2 \leq 2\log\frac{r^2|f'(z)|}{(1 - r^2)|f(z)|^2}$$

$$\leq -2\log\left\{\frac{(1-r)^2}{r}|f(z)|\right\},$$

where $r = |z| < 1$. Let $z = re^{i\theta_0}$, where $e^{i\theta_0}$ is the Hayman direction, then

$$\sum_{n=1}^{\infty} n\left|A_n(re^{i\theta_0}) - \frac{1}{n}r^n e^{-in\theta_0}\right|^2 < -2\log\alpha, \quad r < 1,$$

by Lemma 1.3.1. From (3.3.9),

$$nA_n(re^{i\theta_0}) + r^{-n}e^{-in\theta_0} = F_n\left(\frac{1}{f(re^{i\theta_0})}\right),$$

where F_n is the Faber polynomial of g. Since $f(re^{i\theta_0}) \to \infty$ as $r \to 1$, we have

$$A_n(e^{i\theta_0}) = \lim_{r \to 1} A_n(re^{i\theta_0}) = \frac{1}{n}F_n(0) - \frac{1}{n}e^{-in\theta_0} = 2\gamma_n - \frac{1}{n}e^{-in\theta_0}.$$

Thus

$$4\sum_{n=1}^{\infty} n\left|\gamma_n - \frac{1}{n}e^{-in\theta_0}\right|^2 \leq -2\log\alpha.$$

This is the Bazilevich Inequality (1.3.9).

Finally, we prove the Hayman regularity theorem (Theorem 1.3.1). The proof was given by Aharonov (D. Aharonov [3]). The main tool of the proof is the Bazilevich inequality (1.3.9).

Lemma 3.3.1.   *Let $A_1, A_2, \cdots$ be a sequence of complex numbers, such that*

$$(3.3.15) \qquad \sum_{k=1}^{\infty} k|A_k|^2 < \infty, \quad \Re\left(\sum_{k=1}^{\infty} A_k r^k\right) = O(1), when \quad r \to 1^-,$$

then for any given $\varepsilon > 0$, we may find a constant $M > 0$, such that

$$\left|\exp\left(\sum_{k=1}^{\infty} A_k r^k e^{ik\theta}\right)\right| \leq M\left|\frac{1 - re^{i\theta}}{1 - r}\right|^{\varepsilon},$$

where $0 \leq r < 1$, $0 \leq \theta < 2\pi$, and M is independent of r and θ.
Proof.   By (3.3.15), we have

$$\left|exp\left(\sum_{k=1}^{\infty} A_k r^k e^{ik\theta}\right)\right| = \exp\left(\Re\sum_{k=1}^{\infty} A_k r^k e^{ik\theta}\right)$$

$$= \exp\left[\left(\Re\sum_{k=1}^{\infty} A_k r^k(e^{ik\theta} - 1)\right) + O(1)\right].$$

We may choose N sufficiently large so that

$$(3.3.16) \qquad 2\sum_{k=N+1}^{\infty} k|A_k|^2 < \varepsilon^2.$$

Moreover, we know

$$\log|1 - r^2 e^{i\theta}| = -\Re \log(1 - r^2 e^{i\theta})^{-1}$$

$$= -\Re \sum_{k=1}^{\infty} \frac{r^{2k} e^{ik\theta}}{k} = -\sum_{k=1}^{\infty} \frac{r^{2k} \cos(k\theta)}{k}.$$

Hence

(3.3.17) $$\log \frac{1}{1 - r^2} - \sum_{k=1}^{\infty} \frac{r^{2k} \cos(k\theta)}{k} = \log \left| \frac{1 - r^2 e^{i\theta}}{1 - r^2} \right|.$$

Using (3.3.16) and (3.3.17), we have

$$\left| \Re \sum_{k=1}^{\infty} A_k r^k (e^{ik\theta} - 1) \right| \leq \left| \sum_{k=1}^{N} A_k r^k (e^{ik\theta} - 1) \right|$$

$$+ \left(\sum_{K=N+1}^{\infty} k|A_K|^2 \right)^{\frac{1}{2}} \left(\sum_{k=N+1}^{\infty} \frac{r^{2k}}{k} |e^{ik\theta} - 1|^2 \right)^{\frac{1}{2}}$$

$$= O(1) + 2^{\frac{1}{2}} \left(\sum_{k=N+1}^{\infty} k|A_k|^2 \right)^{\frac{1}{2}} \left(\sum_{k=N+1}^{\infty} \frac{r^{2k}}{k} 2\sin^2 \left(\frac{k\theta}{2} \right) \right)^{\frac{1}{2}}$$

$$< O(1) + \varepsilon \left(\log \left| \frac{1 - r^2 e^{i\theta}}{1 - r^2} \right| \right)^{\frac{1}{2}}$$

$$< O(1) + \varepsilon \left(\log \left| \frac{1 - r e^{i\theta}}{1 - r} \right| \right)^{\frac{1}{2}}.$$

Thus

$$\left| \exp \left(\sum_{k=1}^{\infty} A_k r^k e^{ik\theta} \right) \right| \leq O(1) \exp \left[\varepsilon \left(\log \left| \frac{1 - r e^{i\theta}}{1 - r} \right| \right)^{\frac{1}{2}} \right]$$

$$\leq M \left| \frac{1 - r e^{i\theta}}{1 - r} \right|^{\varepsilon},$$

which completes the proof.

Lemma 3.3.2. *Let $A_1, A_2, \cdots$ be a sequence which satisfies the conditions of Lemma 3.3.1. Let*

$$g(z) = \exp \left(\sum_{k=1}^{\infty} A_k z^k \right),$$

then for any given $\varepsilon, \eta > 0$, we may find $T > 0$, which is independent of z, such that

$$|g(z) - g(r)| \leq T \left| \frac{1 - z}{1 - r} \right|^\varepsilon |e^{i\theta} - 1| \left(1 + \frac{\eta}{1 - r} \right)$$

where $z = re^{i\theta} \in D$, the unit disc.

Proof.   We note

$$g'(z) = \left(\sum_{k=1}^\infty k A_k z^{k-1} \right) \exp \left(\sum_{k=1}^\infty A_k z^k \right).$$

We may choose N sufficiently large, so that

$$2 \sum_{k=N+1}^\infty k |A_k|^2 < \varepsilon^2, \quad \text{and} \quad \sum_{K=N+1}^\infty k |A_k|^2 < \eta^2$$

hold.

Then

$$\left| \sum_{k=1}^\infty k A_k z^{k-1} \right| \leq \sum_{k=1}^N k |A_k| + \left(\sum_{k=N+1}^\infty k |A_k|^2 \right)^{\frac{1}{2}} \left(\sum_{k=N+1}^\infty k r^{2k-2} \right)$$
$$< O(1) + \frac{\eta}{1 - r}.$$

The Lemma follows from the mean value theorem applied to $g(re^{i\theta}) - g(r)$ and using Lemma 3.3.1.

Lemma 3.3.3.   *Let $f(z) \in S$, and let*

$$\lim_{r \to 1} (1 - r)^2 M_\infty(r, f) = \alpha > 0.$$

If the Hayman direction of f is the positive real axis, then for any given $\varepsilon > 0$, we may find $M > 0$, which is independent on $z = re^{i\theta}$, such that

$$|f(z)| < \frac{M}{|1 - z|^{2-\varepsilon}(1 - r)^\varepsilon}$$

holds.

Proof. Let $A_k = 2\left(\gamma_k - \frac{1}{k}\right)$, where γ_k is defined by (1.3.8), then

$$(3.3.18) \qquad f(z) = K(z) \exp\left(\sum_{k=1}^{\infty} A_k z^k\right),$$

where $K(z)$ is the Koebe function (1.1.6). By the Bazilevich inequality (1.3.9), we know $\sum_{k=1}^{\infty} k|A_k|^2 < \infty$, and by Lemma 1.3.2, we know $\Re e\left(\sum_{k=1}^{\infty} A_k r^k\right) = \log\left|\frac{f(r)}{r}(1-r)^2\right| \to \log\alpha$ when $r \to 1^-$. Thus the conditions of Lemma 3.3.1 are satisfied. Using Lemma 3.3.1 to (3.3.18), we have proved Lemma 3.3.3.

Now we are going to prove the Hayman regularity theorem.

Let

$$f_n(z) = K(z)\frac{f(r_n)}{r_n}(1-r_n)^2, \quad r_n = 1 - \frac{1}{n},$$

then the n-th coefficient of the Taylor expansion of f_n, we denote by

$$a_n(f_n) = n\left[\frac{f(r_n)}{r_n}(1-r_n)^2\right].$$

We already assumed that $\alpha > 0$, and the Hayman direction is the positive real axis, we have

$$\frac{a_n(f_n)}{n} \to \alpha, \quad \text{when } n \to \infty.$$

We only need to prove

$$\frac{a_n(f_n - f)}{n} \to 0, \quad \text{when } n \to \infty.$$

We have

$$h_n(z) = f(z) - f_n(z) = K(z)\left[\exp\left(\sum_{k=1}^{\infty} A_k z^k\right) - \exp\left(\sum_{k=1}^{\infty} A_k r_n^k\right)\right].$$

By the Cauchy formula

$$a_n(h_n(z)) = \frac{1}{2\pi i}\int_{|z|=r_n} \frac{h_n(z)}{z^{n+1}} dz$$

$$= \frac{1}{2\pi i}\left(\int_{c_1} + \int_{c_2}\right) \frac{h_n(z)}{z^{n+1}} dz = I_1 + I_2,$$

where

$$c_1 = \{\theta | L(1 - r_n) \leq \theta \leq \pi, \ |z| = r_n\},$$
$$c_2 = \{\theta | \ |\theta| < L(1 - r_n)\},$$

and L is a positive number that will be chosen later.

At c_1, we use $|h_n(z)| \leq |f(z)| + |f_n(z)|$, and Lemma 3.3.3 for $\varepsilon = \frac{1}{2}$, then we have

$$|I_1| \leq \frac{1}{2\pi} \int_{c_1} \frac{|h_n(z)|}{|z|^{n+1}} |dz| \leq \frac{1}{2\pi} \cdot \frac{2M}{(1 - \frac{1}{n})^n} \cdot \frac{1}{(1 - r_n)^{\frac{1}{2}}} \int_{c_1} \frac{d\theta}{|1 - z|^{3/2}}$$

(3.3.19)

$$\leq \frac{M_1 \cdot n}{L^{\frac{1}{2}}},$$

where M_1 is a constant.

At c_2, we use $h_n(Z) = K(r_n e^{i\theta})(g(r_n e^{i\theta}) - g(r e^{i\theta}))$ and Lemma 3.3.2, then we have

$$|h_n(r_n e^{i\theta})| \leq T \frac{r_n}{(1 - r_n)^2} \left| \frac{1 - r_n e^{i\theta}}{1 - r_n} \right| |e^{i\theta} - 1| \left(1 + \frac{\eta}{1 - r_n}\right),$$

and

(3.3.20) $$|I_2| \leq \frac{1}{2\pi} \int_{c_2} \frac{|h_n(z)|}{|z|^{n+1}} |dz| < T_1 L^3 (1 + \eta n),$$

where T_1 is a constant.

We first choose L large enough to make $\dfrac{|I_1|}{n}$ arbitrarily small. This is possible by (3.3.19). Then we choose η such that $T_1 \eta L^3$ is small enough to make $\dfrac{|I_2|}{n}$ arbitrarily small for some large n. This is possible by (3.3.20). This completes the proof.

Here we have proved the Hayman regularity theorem when $\alpha > 0$. When $\alpha = 0$ (cf. Hayman [1]), then for any arbitrary small $\epsilon > 0$, there exists r_0, such that $r_0 < 1$, and

$$M_\infty(r, f) < \frac{\epsilon^2 r}{(1 - r)^2}$$

when $r_0^2 < r < 1$. We deduce that

$$M_\infty(r, h) < \frac{\epsilon r}{1 - r^2}$$

when $r_0 < r < 1$, where $h(z) = \sqrt{f(z^2)}$ with the expansion (1.2.11). Hence

$$\sum_{n=1}^{\infty} n|c_n|^2 r^{2n} < \left(\frac{\epsilon r}{1-r^2}\right)^2,$$

when $r_0 < r < 1$, and

$$M_1(r^2, f) = \sum_{n=1}^{\infty} |c_n|^2 r^{2n} = \int_0^r \sum_1^{\infty} 2n|c_n|^2 \rho^{2n-1} d\rho$$

$$= \int_0^{r_0} + \int_{r_0}^r < \epsilon^2 \int_0^r \frac{2\rho d\rho}{(1-\rho^2)^2} + O(1) = \frac{\epsilon^2 r^2}{1-r^2} + O(1).$$

Choosing $r = 1 - \frac{1}{n}$, we obtain for large n,

$$|a_n| < \frac{1}{r^{n-1}} M_1(r, f) < 2\epsilon^2 en.$$

Thus $\frac{|a_n|}{n} \to 0$.

§3.4 Two applications

In the last section, we have proved the important Milin Theorem, Milin Lemma, Bazilevich Inequality and Hayman Regularity Theorem by the exponentiation of Grunsky Inequality. In this section, we give two more applications of them to coefficient problems. One is the estimate of the difference of the modulus of two successive coefficients, and the another concerns the estimate of the coefficients of odd univalent functions.

Let $f(z) \subset S$, with the expansion (1.1.5). The difference of the modulus of two successive coefficients is $d_n = ||a_{n+1}| - |a_n||$, $n = 2, 3, \cdots$. In 1946, Golusin (G.M.Golusin[7]) proved $d_n \leq Cn^{\frac{1}{4}} \log n$, where C is a constant. In 1956, Biernack(M. Biernack[1]) improved it to $d_n \leq C(\log n)^{\frac{3}{2}}$. Finally, Hayman(W. K. Hayman [3]) proved that the upper bound of d_n was an absolute constant A. In 1968, Milin(E.M. Milin[4]), then Ilina(L.P. Ilina[1]) gave the estimate of the upper bound of A as 4.26. After that the estimate of the upper bound of A was improved again and again. For example, Grinspan(A.Z. Grinspan[1]) proved that:

$$-2.97 < |a_{n+1}| - |a_n| < 3.61.$$

Z. Q. Ye [1] modified Grinspan's proof to improve it to $-2.945 < |a_{n+1}| - |a_n| < 3.394$ in 1985. In 1989, Ke Hu (K. Hu [2]) modified it again to obtain $-2.794 < |a_{n+1}| - |a_n| < 3.26$.

The precise upper and lower bounds of $|a_{n+1}| - |a_n|$ are an open problem. We only know the precise bound in the case $n = 2$ (Theorem 2.3.4) $-1 \leq |a_3| - |a_2| \leq 1.029 \cdots$.

Hayman wrote a very interesting chapter about the difference of successive coefficients for mean p-valent functions in his book [1]. Moreover, Duren [2] proved that for each $f \in S$ with Hayman index $\alpha > 0$,

$$\left||a_{n+1}| - |a_n|\right| \leq e^\delta \alpha^{-\frac{1}{2}} < 1.37 \alpha^{\frac{-1}{2}},$$

where δ is Milin's constant.

For functions with large index α, Duren's work improves (3.4.1).

Pommerenke [4] conjectured that

$$\left||a_{n+1}| - |a_n|\right| \leq 1$$

when $f \in S$ and starlike. This was proved by Leung [1].

Theorem 3.4.1.   *If $f(z) \in S$ and has the expansion (1.1.5), then*

(3.4.1) $-2.794 < |a_{n+1}| - |a_n| < 3.26.$

Let $f(z) = z + \sum\limits_{n=2}^{\infty} a_n z^n \in S$, then $g(w) = \frac{1}{f(z)} \in \Sigma$, where $w = \frac{1}{z}$.

Fixed $\rho \in (0, 1)$, let ζ be a point on the circle $|z| = \rho$ such that $|f(\zeta)|$ takes the maximum value of $|f(z)|$ on the circle.

Let

(3.4.2) $\varphi(z) = \log \dfrac{\frac{1}{z} - \frac{1}{\zeta}}{g\left(\frac{1}{z}\right) - g\left(\frac{1}{\zeta}\right)} = \sum\limits_{n=1}^{\infty} \alpha_n(\zeta) z^n, \quad |\zeta| < 1, |z| < 1,$

(3.4.3) $e^{\varphi(z)} = \sum\limits_{n=0}^{\infty} \beta_n(\zeta) z^n,$

(3.4.4) $x^2 = \dfrac{1}{n} \sum\limits_{k=1}^{n} k^2 |\alpha_k|^2$

and let D_n be the coefficient of z^n in the Taylor expansion of the function $\frac{f(z)}{f(\zeta)} e^{\varphi(z)}$ at $z = 0$.

Lemma 3.4.1 *The following inequalities*

$$(3.4.5) \qquad \sum_{k=0}^{n-1} |\beta_k|^2 \le \frac{n}{(1-\rho^2)(n+\frac{1}{2})} exp(1 - x^2 - \gamma)$$

and

$$(3.4.6) \qquad |\beta_n|^2 \le \frac{x^2}{(1-\rho^2)(n+\frac{1}{2})} exp(1 - x^2 - \gamma)$$

hold, where $\gamma = 0.577\cdots$ is Euler's constant.

Proof. By the Second Lebedev-Milin inequality (1.3.6), we have

$$\sum_{k=0}^{n-1} |\beta_k|^2 \le n \exp\left\{ \frac{1}{n} \sum_{m=1}^{n-1} \sum_{k=1}^{m} (k|\alpha_k|^2 - \frac{1}{k}) \right\}$$

$$= n \exp\left\{ \sum_{k=1}^{n} \left(1 - \frac{k}{n}\right) k|\alpha_k|^2 - \sum_{k-1}^{1} +1 \right\}$$

$$(3.4.7) \qquad = n \exp\left\{ \sum_{k=1}^{n} k|\alpha_k|^2 - x^2 - \sum_{k=1}^{n} \frac{1}{k} + 1 \right\}.$$

From (3.3.6), we know

$$(3.4.8) \qquad \sum_{n=1}^{(\infty)} k|\alpha_k|^2 \le -\log(1-\rho^2).$$

Applying (3.4.8) and the elementary inequalities

$$\log\left(n + \frac{1}{2}\right) < \sum_{k=1}^{n} \frac{1}{k} - \gamma$$

to the righthand side of (3.4.7), we have (3.4.5).
    (3.4.6) is the consequence of (3.3.4) and (3.4.5).

Lemma 3.4.2 *The inequality*

$$(3.4.9) \qquad |D_n| \le \frac{x + \rho^{-n}}{(1-\rho^2)^{\frac{1}{2}}} \left(n + \frac{1}{2}\right)^{\frac{1}{2}} \exp\left\{ \frac{1}{2}(1 - x^2 - \gamma) \right\}$$

holds.

Proof.   By the definition of B_n,

$$D_n = \frac{1}{f(\zeta)} \sum_{k=0}^{n-1} \beta_k \alpha_{n-k}$$

$$= \frac{1}{nf(\zeta)} \left\{ \sum_{k=1}^{n-1} S_k \beta_k \alpha_{n-k} + \sum_{k=1}^{n-1} t_{n-k} \beta_{n-k} \alpha_k \right\}$$

(3.4.10) $$= \frac{1}{nf(\zeta)} \{ I_1 + I_2 \}$$

where $S_k = k\rho^{n-k}, t_k = n - S_k, k = 0, 1, 2, \cdots, n$, and

$$I_1 = \sum_{k=1}^{n-1} S_k \beta_k \alpha_{n-k}, I_2 = \sum_{k=1}^{n-1} t_{n-k} \beta_{n-k} \alpha_k.$$

By (3.4.2) and (3.4.3), we know (3.3.1) holds.

Let $\xi = e^{i\theta}$, then

$$\frac{1}{n|f(\zeta)|}|I_1| = \frac{1}{n|f(\zeta)|} \left| \frac{1}{2\pi} \int_0^{2\pi} \xi^{-2n} \sum_{k=1}^n k\beta_k \xi^k \sum_{l=1}^n \alpha_l \rho^l \xi^l d\theta \right|$$

$$= \frac{1}{n|f(\zeta)|} \left| \frac{1}{2\pi} \int_0^{2\pi} \xi^{-2n} \sum_{k=1}^n k\alpha_k \xi^k \sum_{m=0}^{n-1} \beta_m \xi^m \sum_{l=1}^\infty a_l \rho^l \xi^l d\theta \right|$$

$$\leq \frac{1}{2\pi n} \int_0^{2\pi} \left| \sum_{k=1}^n k\alpha_k \xi^k \right| \left| \sum_{m=0}^{n-1} \beta_m \xi^m \right| d\theta$$

$$\leq \frac{1}{2\pi n} \left[\int_0^{2\pi} \left| \sum_{k=1}^n k\alpha_k \xi^k \right|^2 d\theta \int_0^{2\pi} \left| \sum_{k=0}^{n-1} \beta_k \xi^k \right|^2 d\theta \right]^{\frac{1}{2}}$$

$$= \frac{1}{n} \left[\sum_{k=1}^n k^2 |\alpha_k|^2 \sum_{k=0}^{n-1} |\beta_k|^2 \right]^{\frac{1}{2}}$$

by Schwarz inequality and $|f(\zeta)| = \max\limits_{|z|=\rho} |f(z)|$.

It follows from (3.4.5) that

(3.4.11) $$\frac{1}{n|f(\zeta|)} |I_1| \leq \frac{x}{(1-\rho^2)^{\frac{1}{2}} \left(n + \frac{1}{2}\right)^{\frac{1}{2}}} \exp \left\{ \frac{1}{2} (1 - x^2 - \gamma) \right\}.$$

It is easily observed that $t_{n-k}\rho^{n-k} \leq k \leq \sqrt{kn}$ when $1 \leq k \leq n$, and
$$\sum_{k=1}^{\infty} k|\alpha_k|^2 \rho^{2k} \leq \left(\max_{|z|=\rho}|f(z)|\right)^2.$$
Thus

$$|I_2| = \left|\sum_{y=1}^{n-1} y_{n-k}\beta_{n-k}\alpha_k\right| \leq \sqrt{n}\rho^{-n}\left|\sum_{k=1}^{n-1}\sqrt{k}\rho^k\alpha_k\beta_{n-k}\right|$$

$$\leq \sqrt{n}\rho^{-n}\left[\sum_{k=1}^{n-1}|\beta_k|^2\sum_{k=1}^{n-1}k|\alpha_k|^2\rho^{2k}\right]^{\frac{1}{2}}$$

$$\leq \sqrt{n}\rho^{-n}|f(\zeta)|\left(\sum_{k=1}^{n-1}|\beta_k|^2\right)^{\frac{1}{2}}.$$

By (3.4.5), the previous inequality becomes

(3.4.12) $$\frac{1}{n|f(\zeta)|}|I_2| \leq \frac{\rho^{-n}}{(1-\rho^2)^{\frac{1}{2}}\left(n+\frac{1}{2}\right)^{\frac{1}{2}}}\exp^{\frac{1}{2}}(1-x^2-\gamma).$$

Inequality (3.4.9) follows from (3.4.10), (3.4.11) and (3.4.12).

Lemma 3.4.3 *If* $|a_{n+1}| > 0, \rho^2 = exp\left(\frac{-1}{n+1}\right)$, *then*

(3.4.13) $$|D_n| \leq \frac{n(n+1)^{\frac{1}{2}}\rho^{-n-1}}{\sqrt{3}|a_{n+1}|^2(1-\rho^2)^{\frac{1}{2}}}\,exp\frac{1}{2}(1-x^2-\gamma)$$

holds.

Proof. By de Branges Theorem (Theorem 4.2.3)

$$\sum_{k=1}^{n}|a_n|^2 \leq \sum_{K=1}^{n}k^2 = \frac{1}{3}n(n+1)(n+\frac{1}{2}).$$

We observe that, if $\rho^2 = \exp\left(\frac{-1}{n+1}\right)$, then

$$(k+n+1)\rho^{2k} > (k+n+1)(1-\frac{k}{n+1}) > n+1-k.$$

It follows from FitzGerald inequality (2.4.15) that

$$|a_{n+1}|^4\rho^{2(n+1)} \leq \left\{\sum_{k=1}^{n+1}k|a_k|^2 + \sum_{k=n+2}^{2n+1}(2n+2-k)|a_k|^2\right\}$$

$$\times \rho^{2(n+1)} \leq \sum_{k=1}^{\infty}k|a_k|^2\rho^{2k} \leq |f(\zeta)|^2.$$

Thus we obtain

$$|D_n| = \frac{1}{|f(\zeta)|} \left| \sum_{k=1}^{n} a_k \beta_{n-k} \right| \le \frac{1}{|f(\zeta)|} \left[\sum_{k=1}^{n} |a_k|^2 \sum_{k=0}^{n-1} |\beta_k|^2 \right]^{\frac{1}{2}}$$

(3.4.14)

$$\le \frac{\{n(n+1)(n+\frac{1}{2})\}^{\frac{1}{2}}}{\sqrt{3}|a_{n+1}|^2 \rho^{n+1}} \left(\sum_{k=0}^{n-1} |\beta_k|^2 \right)^{\frac{1}{2}}.$$

Using (3.4.5) to the righthand side of (3.4.14), we complete the proof of (3.4.13).

   Proof Theorem 3.4.1.

   If $f \in S$, we have the following identity

$$\left(\frac{1}{z} - \frac{1}{\zeta} \right) f(z) = \left\{ 1 - \frac{f(z)}{f(\zeta)} \right\} \frac{\frac{1}{z} - \frac{1}{\zeta}}{g\left(\frac{1}{z}\right) - g\left(\frac{1}{\zeta}\right)} = e^{\varphi(z)} - \frac{f(z)}{f(\zeta)} e^{\varphi(z)}$$

where $g\left(\frac{1}{z}\right) = \frac{1}{f(z)}, |f(\zeta)| = \max\limits_{|z|=\rho} |f(z)|.$

   From this identity, we have

$$a_{n+1} - \zeta^{-1} a_n = \beta_n - D_n.$$

Thus

$$\left| |a_{n+1}| - \rho^{-1} |a_n| \right| \le |a_{n+1} - \zeta^{-1} a_n| \le |\beta_n| + |D_n|$$

$$\le \frac{zx + \rho^{-n}}{(1-\rho^2)^{\frac{1}{2}} (n+\frac{1}{2})^{\frac{1}{2}}} \exp \frac{1}{2} (1 - x^2 - \gamma)$$

$$\le \frac{1}{(1-\rho^2)^{\frac{1}{2}} (n+\frac{1}{2})^{\frac{1}{2}}} \frac{\rho^{-n} + (\rho^{-2n} + 16)^{\frac{1}{2}}}{2}$$

$$\times \exp \frac{1}{2} \left\{ 1 - \gamma - \left(\frac{(\rho^{-2n} + 16)^{\frac{1}{2}} - \rho^{-n}}{4} \right)^2 \right\}$$

(3.4.15)

$$\equiv G(\rho).$$

Let $\rho = \exp\left(\frac{-1}{1.3n}\right)$ and performing some computation, we conclude that

$$\left| |a_{n+1}| - \rho^{-1} |a_n| \right| \le G\left(\exp\left(\frac{1}{1.3n} \right) \right) < 2.79375.$$

It implies
$$|a_{n+1}| - |a_n| > |a_{n+1}| - \rho^{-1}|a_n| > -2.79375.$$

We have proved the lefthand side inequality of (3.4.1).

To prove the righthand side inequality of (3.4.1), we consider it in the following two cases.

(1) If $|a_{n+1}| \le 0.6685(n+1)$.

Let $\rho = \exp\left(\frac{-1}{1.5(n+1)}\right)$, then

$$
|a_{n+1}| - |a_n| = (1-\rho)|a_{n+1}| + \rho|a_{n+1}| - |a_n|
$$
$$
\le \left(1 - \exp\left(\frac{-1}{1.5(n+1)}\right)\right)|a_{n+1}| + \exp\left(\frac{-1}{1.5(n+1)}\right)
$$
$$
\times G\left(\left(\frac{-1}{1.5(n+1)}\right)\right) < 3.26
$$

by (3.4.15).

(2) If $|a_{n+1}| > 0.6685(n+1)$.

We have

$$
\left||a_{n+1}| - \rho^{-1}|u_n|\right| \le |\beta_n| + |D_n|
$$
$$
\le \frac{1}{(1-\rho^2)^{\frac{1}{2}}\left(n+\frac{1}{2}\right)^{\frac{1}{2}}}\left\{x + \frac{n(n+1)^{\frac{1}{2}}\left(n+\frac{1}{2}\right)^{\frac{1}{2}}}{\sqrt{3}|a_{n+1}|^2\rho^{n+1}}\right\}\exp\frac{1}{2}(1 - x^2 - \gamma)
$$
$$
\equiv H(\rho, |a_{n+1}|).
$$

Let $\rho^2 = \exp\left(\frac{1}{n+1}\right), |a_{n+1}| = y(n+1), \ (0.6685 < y \le 1)$, then

$$
|a_{n+1}| - |a_n| = \rho|a_{n+1}| - |a_n| + (1-\rho)|a_{n+1}|
$$
$$
\le \rho H(\rho, |a_{n+1}|) + (1-\rho)y(n+1)
$$
$$
\le \frac{1}{2}y + \frac{1}{2}\left\{\frac{\sqrt{e}}{\sqrt{3}y^2} + \left(\frac{e}{3y^4} + 4\right)^{\frac{1}{2}}\right\}
$$
$$
\times \exp\frac{1}{2}\left\{1 - C - \frac{1}{2}\left(\left(\frac{e}{3y^4} + 4\right)^{\frac{1}{2}} - \frac{\sqrt{e}}{\sqrt{3}y^2}\right)^2\right\}
$$
$$
\equiv J(y) \le J(0.6685) \le 3.26.
$$

We have proved the righthand side inequality of (3.4.1).

Next, we will state the estimate of the modulus of the coefficients of odd univalent functions in S. This result was proved by Ke Hu (K. Hu [1]) by modifying the proof of Milin's result.

Of course, the precise bound is an open problem.

Theorem 3.4.2   *Let* $h(z) = z + \sum\limits_{n=1}^{\infty} c_{2n+1} z^{2n+1} \in S$, *then*

$$|c_{2n+1}| < 1.1305$$

for $n = 2, 3, \cdots$.

Proof.   de Branges proved the Milin conjecture (Theorem 4.2.3). If

$$(3.4.16) \qquad\qquad \log \frac{h(\sqrt{z})}{\sqrt{z}} = \sum_{n=1}^{\infty} \gamma_n z^n,$$

then

$$(3.4.17) \qquad\qquad \sum_{k=1}^{n}(n+1-k)k|\gamma_k|^2 \le \sum_{k=1}^{n}\frac{n+1-k}{k}, \quad n = 2, 3, \cdots$$

holds.

For any $\rho, 0 < \rho < 1$, we have

$$\sum_{k=1}^{n} k|\gamma_k|^2 \rho^k = (1-\rho)^2 \left(\sum_{k=1}^{n} k|\gamma_k|^2 \rho^k \right) \left(\sum_{k=0}^{\infty}(k+1)\rho^k \right)$$

$$= (1-\rho)^2 \left\{ \sum_{k=1}^{n} \left[\sum_{i=1}^{k}(k-i+1)i|\gamma_i|^2 \right] \rho^k \right.$$

$$+ \rho^n \sum_{\nu=1}^{\infty} \left[\sum_{k=1}^{n}(n+\nu-i+1)i|\gamma_i^2| \right] \rho^\nu \bigg\}$$

$$\le (1-\rho)^2 \left\{ \sum_{k=1}^{n} \left(\sum_{i=1}^{k}\frac{k-i+1}{i} \right) \rho^k + \frac{\rho^{n+1}}{1-\rho} \sum_{i=1}^{n}\frac{n-i+1}{i} \right.$$

$$(3.4.18)$$

$$+ \frac{\rho^{n+1}}{(1-\rho)^2} \sum_{i=1}^{n} i|\gamma_i|^2 \bigg\}$$

by (3.4.17).

By the equality

$$\sum_{k=1}^{n} \sum_{i=1}^{k} \frac{k-i+1}{i}\rho^k = \sum_{i=1}^{n}\sum_{k=i}^{n}\frac{k-i+1}{i}\rho^k = \sum_{i=1}^{n}\frac{\rho^i}{i}\sum_{l=1}^{n-i+1} l\rho^{l-1}$$

$$= \sum_{i=1}^{n}\frac{\rho^i}{i}\frac{1-(n-i+2)\rho^{n-i+1}+(n-i+1)\rho^{n-i+2}}{(1-\rho)^2},$$

the righthand side of inequality (3.4.18) is equal to

$$\sum_{k=1}^{n} \frac{1}{k}\rho^k + \rho^{n+1}\sum_{i=1}^{n}\left(i|\gamma_i|^2 - \frac{1}{i}\right)$$

which is not greater than

$$\sum_{k=1}^{n}\frac{1}{k}\rho^k + \rho^{n+1}\delta$$

by Milin Lemma (1.3.10), where $\delta < 0.312$ is the Milin constant.
Hence, we have

(3.4.19)
$$\sum_{k=1}^{n} k|\gamma_k|^2\rho^k \leq \sum_{k=1}^{n}\frac{\rho^k}{k} + \rho^{n+1}\delta.$$

We apply the Third Lebedev-Milin Inequality (1.3.7) to (3.4.16). We obtain the inequality

(3.4.20)
$$|c_{2n+1}| \leq \exp\frac{1}{2}\left\{\sum_{k=1}^{n}\left(k|\gamma_k|^2\rho^k - \frac{1}{k}\right) - n\log\rho\right\}$$

for any $0 < \rho < 1$.

By (3.4.19) and (3.4.20), we have the inequality

(3.4.21)
$$|c_{2n+1}| \leq \exp\frac{1}{2}\left\{\sum_{k=1}^{n}\frac{\rho^k}{k} + \rho^{n+1}\delta - \sum_{k=1}^{n}\frac{1}{k} - n\log\rho\right\}$$

for any $0 < \rho < 1$.

Let $\rho = \exp\left\{\frac{-2y}{2n+1}\right\}$, $y > 0$, then

$$\sum_{k=n+1}^{\infty}\frac{\rho^k}{k} = \frac{2}{2n+1}\int_{y}^{\infty}\frac{e^{-t}dt}{e^{\frac{t}{2n+1}} - e^{\frac{-t}{2n+1}}}.$$

Applying the following inequality

$$\frac{1}{e^x - e^{-x}} \geq \frac{1}{2x}e^{-x^2/6} > \frac{1}{2x}\left(1 - \frac{x^2}{6}\right), \quad (x > 0)$$

to the previous equality, we have
(3.4.22)
$$\sum_{k=n+1}^{\infty} \frac{\rho^k}{k} > \int_y^{\infty} \frac{e^{-t}}{t} \left(1 - \frac{t^2}{6(2n+1)^2}\right) dt = \int_y^{\infty} \frac{e^{-t}dt}{t} - \frac{e^{-y}(1+y)}{6(2n+1)^2}.$$

Moreover, the inequality

$$\frac{1}{e^x - e^{-x}} \leq \frac{1}{2x(1 + \frac{1}{3}x^2)},$$

and the equality

$$\log(1+x) = \left\{\frac{x}{1+x} + \frac{1}{2}\left(\frac{x}{1+x}\right)^2 + \cdots\right\}$$

hold when $x > 0$. Using these results, we have

$$\log\frac{1}{1-\rho} = \log\frac{e^{\frac{y}{2n+1}}}{e^{\frac{y}{2n+1}} - e^{\frac{-y}{2n+1}}}$$

$$< \frac{y}{2n+1} + \log\left\{\frac{2y}{2n+1}\left[1 + \frac{1}{3}\left(\frac{y}{2n+1}\right)^2\right]\right\}^{-1}$$

$$\leq \frac{y}{2n+1} + \log\frac{1}{y} + \log\left(n + \frac{1}{2}\right) - \frac{y^2}{3(2n+1)^2 + y^2}$$

(3.4.23)
$$\leq \sum_{k=1}^{n}\frac{1}{k} - \gamma + \log\frac{1}{y} + \frac{y}{2n+1} - \frac{y^2}{3(2n+1)^2 + y^2}$$

where the inequality $\log\left(n + \frac{1}{2}\right) < \sum_{k=1}^{n}\frac{1}{k} - \gamma$ has been used, and γ is the Euler constant.

Inequalities (3.4.22) and (3.4.23) lead to the estimate

$$\sum_{k=1}^{n}\frac{\rho^k}{k} = \log\frac{1}{1-\rho} - \sum_{k=n+1}^{\infty}\frac{\rho^k}{k} \leq \sum_{k=1}^{n}\frac{1}{k} - \gamma - \int_y^{\infty}\frac{e^{-t}}{t}dt + \log\frac{1}{y}$$

$$+ \frac{y}{2n+1} + \frac{e^{-y}(1+y)}{6(2n+1)^2} - \frac{y^2}{3(2n+1)^2 + y^2}$$

where $\rho = \exp\left\{\frac{-2y}{2n+1}\right\}$.

Hence

$$\sum_{k=1}^{n} \frac{\rho^k}{k} + \delta\rho^{n+1} - n\log\rho - \sum_{k=1}^{n} \frac{1}{k}$$

$$\leq y + \log\frac{1}{y} - \gamma - \int_{y}^{\infty} \frac{e^{-t}dt}{t} + \frac{e^{-y}(1+y)}{6(2n+1)^2} - \frac{y^2}{3(2n+1)^2 + y^2}$$

$$+ \delta e^{-y-\frac{y}{2n+1}}.$$

Let $y = 0.4574$, then the right-hand side of the previous inequality is less than

$$y + \log\frac{1}{y} - \gamma - \int_{y}^{\infty} \frac{e^{-t}dt}{t} + 0.312e^{-y}$$

when $n \geq 3$. It is equal to

$$0.4574 - 0.577215 + 0.167745 + 0.197473 = 0.245433\cdots$$

Substituting this estimate to (3.4.21), we have

$$|c_{2n+1}| \leq e^{0.1227\cdots} < 1.1305$$

It is known that $|c_3| \leq 1, |c_5| \leq 1.013\cdots$. We have proved the Theorem.

In 1970, Aharonov(D.Aharonov[1]) proved the following theorem: Let $h(z) = z + \sum_{n-1}^{\infty} c_{2n+1}z^{2n+1} \in S$, then $|c_{2n+1}| < 1$ for $n = 2, 3, \cdots$ if $|c_3| < 0.4335$.

Proof. From (3.3.11),

$$2\sum_{k=2}^{n} k|\gamma_k|^2 \leq \sum_{k=2}^{n} \frac{1}{k}\rho^{2k} - \log(1-\rho^{-2}) < \sum_{k=1}^{n} \rho^{2k} - \log(1-\rho^{-2}) - 1.$$

Using a similar argument to the proof of the Milin Lemma, to find the value of ρ, such that the right hand side of the previous formula is minimized, we get

$$\sum_{k=1}^{n} k|\gamma_k|^2 < \sum_{k=1}^{n} \frac{1}{k} + \delta - \frac{1}{2} + \frac{1}{4}|a_2|^2$$

since $2\gamma_1 = a_2$. By (1.3.12), we get

$$|c_{2n+1}|^2 < \exp\left(\delta - \frac{1}{2} + \frac{1}{4}|a_2|^2\right).$$

The result $|c_{2n+1}| < 1$ when $|a_2| < 0.867$ follows because $\delta < 0.312$. This implies $|c_{2n+1}| < 1$ holds for all $n = 2, 3, \cdots$ when $|c_3| < 0.4335$.

After that, there were many improvements.   In 1973, Aharonov
(D. Aharonov[2]), and then Ilina(L. P. Ilina[2]) improved the number 0.4335
to 0.525 independently.

Of course, it implies the following result. If $f(z) \in S$, and has the expan-
sion (1.1.5), then $|a_n| < n$ for $n = 2, 3, \cdots$, when $|a_2| < 1.05$.

In 1974, Ehrig(G. Ehrig[1],[2]) improved the number 1.05 to 1.15. Based
upon the method of Ehrig, Bishouty(D. H. Bishouty[1]) improved it to 1.55
in 1976, and then he improved it to 1.59 in 1983(D. H. Bishouty[2]). But
Gong(S. Gong[5]) already improved it to 1.635 in 1978. Obviously, it is
impossible to prove the Bieberbach conjecture this way.

CHAPTER IV

DE BRANGES THEOREM

$4.1. Askey-Gasper theorem

De Branges proved the Bieberbach Conjecture in 1984 (de Branges [1], [2],[3]). Using the Löwner Theory he proved the Milin conjecture. In his proof, he also used the positivity of the sums of Jacobi polynomials – a result obtained by Askey-Gasper (R. Askey and G. Gasper [1]) in 1976. A key step in the proof of the Askey-Gasper Theorem is the Gegenbauer-Hua formula (L. K. Hua [1]). In this section, we give a detailed proof of these special function results, in preparation for a proof of the de Branges Theorem.

Jacobi Polynomials　For $\alpha > -1$ and $\beta > -1$, $\{P_n^{(\alpha,\beta)}(x)\}$ are polynomials of degree n satisfying the following orthogonality condition

$$(4.1.1) \qquad \int_{-1}^{1} P_n^{(\alpha,\beta)}(x) P_m^{(\alpha,\beta)}(x)(1-x)^{\alpha}(1+x)^{\beta}\, dx = \delta_{n,m},$$

and the normalization condition

$$P_n^{(\alpha,\beta)}(1) = \binom{n+\alpha}{\alpha}.$$

where $\delta_{n,m}$ is the Kronecker delta.

Jacobi polynomials have the following simple properties:

Let f and g are two functions of x, we define the inner product of f and g by

$$(f,g) = \int_{-1}^{1} f(x)g(x)(1-x)^{\alpha}(1+x)^{\beta} dx.$$

If $(f,g) = 0$, we say that f and g are orthogoral, and denote this by $f \perp g$.

105

1. *For a fixed $\alpha > -1$, and $\beta > -1$, if $\pi_n = \sum_{k=0}^{n} a_k x^k$ is an arbitrary polynomial orthogonal to $1, x, x^2, \ldots, x^{n-1}$, then $\pi_n = c \cdot P_n^{(\alpha,\beta)}(x)$, where c is a constant.*

Proof. We rewrite $\pi_n = \sum_{k=0}^{n} a_k x^k$ as $\sum_{k=0}^{n} c_k P_k^{(\alpha,\beta)}(x)$, then $\pi_n \perp P_0^{(\alpha,\beta)}(x), \ldots, P_{n-1}^{(\alpha,\beta)}(x)$ since $\pi_n \perp 1, x, \ldots, x^{n-1}$. Thus, $(\pi_n, P_j) = c_j = 0$, when $0 \le j \le n-1$, and $\pi_n = c \cdot P_n^{(\alpha,\beta)}(x)$.

2. *$P_n^{(\alpha,\beta)}(x)$ satisfies the following differential equation:*

$$(4.1.2) \quad (1-x^2)y'' + [(\beta - \alpha) - (\alpha + \beta + 2)x]y' + n(n + \alpha + \beta + 1)y = 0.$$

Proof. (4.1.2) is equivalent to

$$(4.1.3) \quad [y'(1-x)^{\alpha+1}(1+x)^{\beta+1}]' = -n(n + \alpha + \beta + 1)y(1-x)^\alpha(1+x)^\beta$$

since

$$[y'(1-x)^{\alpha+1}(1+x)^{\beta+1}]'$$
$$= (1-x)^\alpha(1+x)^\beta[y''(1-x^2) + y'((\beta - \alpha) - (\alpha + \beta + 2)x)].$$

Substitution of $y = P_n^{(\alpha,\beta)}(x)$ into the left hand side of (4.1.3) yields an expression of the form $(1-x)^\alpha(1+x)^\beta z$. Obviously, z is a polynomial of degree n. If we can prove $z = -n(n + \alpha + \beta + 1)P_n^{(\alpha,\beta)}(x)$, then we have proved that $P_n^{(\alpha,\beta)}(x)$ satisfies (4.1.2).

First, we prove $z \perp P_j^{(\alpha,\beta)}(x)$, $0 \le j \le n-1$.
For $\alpha > -1$, $\beta > -1$, we have

$$\int_{-1}^{1} (1-x)^\alpha(1+x)^\beta x^j z(x)dx = \int_{-1}^{1} [y'(1-x)^{\alpha+1}(1+x)^{\beta+1}]'x^j dx,$$

Using integration by parts on the right hand side , we obtain

$$-\int_{-1}^{1} y'(1-x)^{\alpha+1}(1+x)^{\beta+1}jx^{j-1}dx,$$

and integrating by parts again, we have

$$j\int_{-1}^{1} y\frac{d}{dx}((1-x)^{\alpha+1}(1+x)^{\beta+1}x^{j-1})dx = \int_{-1}^{1} y(1-x)^\alpha(1+x)^\beta \pi_j(x)dx$$

where $\pi_j(x)$ is a polynomial of degree j. Since $P_n^{(\alpha,\beta)}(x)$ is orthogonal to $1, x, \ldots, x^{n-1}$, the integral on the right side of the previous equation equals zero when $0 \le j \le n-1$, that is, $z \perp x^j$, $0 \le j \le n-1$. By **1**, we have $z(x) = cP_n^{(\alpha,\beta)}(x)$, where c is a constant. Now we determine the constant c. Since

$$\left[\frac{d}{dx}P_n^{(\alpha,\beta)}(x)(1-x)^{\alpha+1}(1+x)^{\beta+1}\right]' = (1-x)^\alpha(1+x)^\beta z$$
$$= (1-x)^\alpha(1+x)^\beta cP_n^{(\alpha,\beta)}(x),$$

we have

$$cP_n^{(\alpha,\beta)}(x) = \frac{d^2}{dx^2}P_n^{(\alpha,\beta)}(x)(1-x^2) + [(\beta-\alpha)-(\alpha+\beta+2)x]\frac{d}{dx}P_n^{(\alpha,\beta)}(x).$$

Letting $P_n^{(\alpha,\beta)}(x) = \sum_{k=0}^n a_k x^k$ in the previous equation, we get $c = -n(n+\alpha+\beta+1)$ when we compare the coefficient of x^n on both sides of previous equation. Thus $z = -n(n+\alpha+\beta+1)P_n^{(\alpha,\beta)}(x)$.

3. *The solution of equation (4.1.2) is unique in the following sense: If y and z are two independent solutions of (4.1.2), then at least one of them is not regular at $x = -1(x = 1)$. In particular, if y and z are polynomial solutions, then they are linearly dependent.*

Proof. In general, if

$$P_2 y'' + P_1 y' + P_0 y = 0, \quad P_2 z'' + P_1 z' + P_0 z = 0,$$

then

$$P_2(y''z - z''y) + P_1(y'z - z'y) = 0,$$

that is,

$$P_2(y'z - z'y)' + P_1(y'z - z'y) = 0.$$

We have

$$y'z - z'y = c\exp\left[-\int^x \frac{P_1}{P_2}dt\right],$$

where c is a constant. Specifically, if we take $P_1 = (\beta-\alpha)-(\alpha+\beta+2)x$, $P_2 = 1-x^2$, then

$$y'z - z'y = c\exp\left[-\int^x \left(\frac{\beta-\alpha}{1-t^2} - \frac{(\alpha+\beta+2)t}{1-t^2}\right)dt\right]$$
$$= c_1(1+x)^{-\beta-1}(1-x)^{-\alpha-1}$$

where c, c_1 are constants, that is,

$$(1+x)^{\beta+1}(1-x)^{\alpha+1}(y'z - z'y) = c_1.$$

If y and z are regular at $x = -1(x = 1)$, then $c_1 = 0$, i.e., $y'z - z'y \equiv 0$. This implies that $y = c_2 z$ where c_2 is a constant.

Jacobi polynomials can be expressed by hypergeometric functions.

A *Hypergeometric function* is defined as follows. For $|t| < 1$, $c \neq 0$, -1, -2, $\cdots$,

$$(4.1.4) \qquad {}_2F_1(a, b; c; t) = F(a, b; c; t) = \sum_{k=0}^{\infty} \frac{(a)_k(b)_k}{(c)_k} \frac{t^k}{k!},$$

$$(4.1.5) \qquad {}_3F_2(a, b, c; d, e; t) = \sum_{k=0}^{\infty} \frac{(a)_k(b)_k(c)_k}{(d)_k(e)_k} \frac{t^k}{k!},$$

$$\cdots\cdots\cdots\cdots$$

where $(m)_k = \frac{\Gamma(m+k)}{\Gamma(m)}$.

It is easy to verify that ${}_2F_1(a, b; c; t)$ satisfies the following differential equation:

$$(4.1.6) \qquad t(1-t)y'' + [c - (a+b+1)t]y' - aby = 0,$$

and ${}_3F_2(a, b, c; d, e; t)$ satisfies the following differential equation:

$$t^2(1-t)z''' - [(3+a+b+a)t^2 - (1+d+e)t]z'' + [de - (1+a+b+c$$
$$(4.1.7)$$
$$+ ab + ac + bc)t]z' - abcz = 0$$

Let $p = t\frac{d}{dt}$ be an operator, then $pt^k = kt^k$. Let $y = {}_2F_1(a, b; c; t)$. Obviously,

$$p(p + c - 1)y = \sum_{k=1}^{\infty} \frac{k(k+c-1)(a)_k(b)_k}{k!(c)_k} t^k = \sum_{k=1}^{\infty} \frac{(a)_k(b)_k t^k}{(k-1)!(c)_{k-1}}$$

$$= \sum_{s=0}^{\infty} \frac{(a)_{s+1}(b)_{s+1}t^{s+1}t^{s+1}}{s!(c)_s} = \sum_{s=0}^{\infty} \frac{(a+s)(b+s)(a)_s(b)_s t^{s+1}}{S!(c)_s}$$

$$= t(p+a)(p+b)y.$$

Thus $_2F_1(a, b; c; t)$ satisfies the differential equation

$$[p(p + c - 1) - t(p + a)(p + b)]y = 0$$

with $y(0) = 1, y'(0) = \frac{ab}{d}$. This is equivalent to saying that $_2F_1(a, b; c; t)$ satisfies differential equation (4.1.6) with $y(0) = 1, y'(0) = \frac{ab}{d}$.

Similarly, let $z = {}_3F_2(a, b, c; d, e; t)$, then we can prove that z satisfies the differential equation

$$[p(p + d - 1)(p + e - 1) - t(p + a)(p + b)(p + c)]z = 0$$

with $z(0) = 1, z'(0) = \frac{abc}{de}$, and $z''(0) = \frac{(a+1)(b+1)(c+1)}{(d+1)(e+1)}$. This is equivalent to saying that $_3F_2(a, b, c; d, e; t)$ satisfies the differential equation (4.1.8) with $z(0) = 1, z'(0) = \frac{abc}{de}$, and $z'''(0) = \frac{(a+1)(b+1)(c+1)}{(d+1)(e+1)}$.

Theorem 4.1.1. *When $\alpha > -1$, $\beta > -1$, $n \geq 1$, then*

$$(4.1.8) \qquad P_n^{(\alpha, \beta)}(x) = \binom{n + \alpha}{n} F\left(-n, n + \alpha + \beta + 1; \alpha + 1; \frac{1 - x}{2}\right).$$

Proof. $F(a, b; c; t)$ satisfies the equation (4.1.6). Using the substitution $t = \frac{1-x}{2}$ in (4.1.6), this becomes

$$(1 - x^2)G''(x) - [2c - (a + b + 1) + (a + b + 1)x]G'(x) - abG(x) = 0$$

where $G(x) = y(t)$. Comparing this equation with (4.1.2), we have

$$ab = -n(n + \alpha + \beta + 1),$$
$$a + b = \alpha + \beta + 1,$$
$$a + b + 1 - 2c = \beta - \alpha.$$

The solutions of the equation

$$(u + a)(u + b) = u^2 + (a + b)u + ab = u^2 + (\alpha + \beta + 1)u - n(n + \alpha + \beta + 1) = 0$$

are

$$u_1 = -n - \alpha - \beta - 1, \quad u_2 = n.$$

We obtain $c = \alpha + 1$ by solving the equation

$$-(\beta - \alpha) + a + b + 1 = 2c = -(\beta - \alpha) + \alpha + \beta + 2 = 2\alpha + 2.$$

$F(a, b; c; t)$ is the solution of (4.1.6), and it is also the solution of (4.1.2) when $a = -n$, $b = n + \alpha + \beta + 1$, and $c = \alpha + 1$. The solution of (4.1.2) is unique, thus $F\left(-n, n + \alpha + \beta + 1; \alpha + 1; \frac{1-x}{2}\right)$ is linearly dependent to $P_n^{(\alpha,\beta)}(x)$, i.e.,

$$F\left(-n, n + \alpha + \beta + 1; \alpha + 1; \frac{1 - x}{2}\right) = cP_n^{(\alpha,\beta)}(x).$$

The constant c is determined by $F(a, b; c; 0) = 1$ and $P_n^{(\alpha,\beta)} = \binom{n+\alpha}{n}$. The Theorem is proved.

Let

$$(4.1.9) \qquad G^{(\lambda)}(x, w) = \frac{1}{(1 - 2xw + w^2)^\lambda} = \sum_{n=0}^{\infty} P_n^{(\lambda)}(x)w^n$$

if $\lambda > -\frac{1}{2}$, $-1 \leq x \leq 1$, and $|w| < 1$. The polynomials $\{P_n^{(\lambda)}\}_{n=0}^{\infty}$ are called *ultraspherical polynomials* or *Gegenbauer polynomials*. Let $x = 1$, then

$$G^{(\lambda)}(1, w) = \frac{1}{(1 - w)^{2\lambda}} = \sum_{n=0}^{\infty} P_n^{(\lambda)}(1)w^n$$

when $|w| < 1$. This implies $P_n^{(\lambda)}(1) = \binom{-2\lambda}{n} = \frac{(2\lambda)_n}{n!}$.

It is easy to verify that $G = G^{(\lambda)}(x, t) = (1 - 2xt + t^2)^{-\lambda}$ satisfies the partial differential equation

$$(4.1.10) \qquad (1 - x^2)G_{xx} - x(2\lambda + 1)G_x + t(tG)_t + (2\lambda t)G_t = 0,$$

and hence the Gegenbauer polynomial $y = P_n^{(\lambda)}(x)$ satisfies the differential equation

$$(4.1.11) \qquad (1 - x^2)y'' - (2\lambda + 1)xy' + n(n + 2\lambda)y = 0.$$

Proof. By the definition of $P_n^{(\lambda)}$, $G = G^{(\lambda)}(x, t) = \sum_{n=0}^{\infty} P_n^{(\lambda)}(x)t^n$, hence

$$G_x = \sum_{n=0}^{\infty} (P_n^{(\lambda)}(x))' t^n, \quad G_{xx} = \sum_{n=0}^{\infty} (P_n^{(\lambda)})'' t^n,$$

$$tG_t = \sum_{n=1}^{\infty} nP_n^{(\lambda)}(x)t^n, \quad t(tG_t)_t = \sum_{n=1}^{\infty} n^2 P_n^{(\lambda)}(x)t^n.$$

Substituting these expression into (4.1.9), we have

$$\sum_{n=0}^{\infty} t^n \{ (1-x^2)(P_n^{(\lambda)}(x))'' - (2\lambda+1)x(P_n^{(\lambda)}(x))'$$
$$+ n^2 P_n^{(\lambda)}(x) + 2\lambda n P_n^{(\lambda)}(x) \} = 0.$$

This proves that $P_n^{(\lambda)}(x)$ satisfies equation (4.1.11).

Theorem 4.1.2. *If* $\lambda > -\frac{1}{2}$, $n \geq 1$, *and* $-1 \leq x \leq 1$, *then*

$$P_n^{(\lambda)}(x) = \frac{(2\lambda)_n}{n!} \frac{P_n^{(\lambda-\frac{1}{2},\lambda-\frac{1}{2})}(x)}{P_n^{(\lambda-\frac{1}{2},\lambda-\frac{1}{2})}(1)}$$

(4.1.12)
$$= \frac{(2\lambda)_n}{n!} \, _2F_1\left(-n, n+2\lambda; \lambda+\frac{1}{2}; \frac{1-x}{2}\right).$$

Proof. Taking $\alpha = \beta = \lambda - \frac{1}{2}$ in Theorem 4.1.1, we find

$$\frac{P_n^{(\lambda-\frac{1}{2},\lambda-\frac{1}{2})}(x)}{P_n^{(\lambda-\frac{1}{2},\lambda-\frac{1}{2})}(1)} = \, _2F_1\left(-n, n+2\lambda; \lambda+\frac{1}{2}; \frac{1-x}{2}\right).$$

Then $P_n^{(\lambda-\frac{1}{2},\lambda-\frac{1}{2})}(x)$ satisfies the equation (4.1.2) when $\alpha = \beta = \lambda - \frac{1}{2}$,

$$(1-x^2)y'' - (2\lambda+1)xy' + n(n+2\lambda)y = 0.$$

This is equation (4.1.10). By 3, equation (4.1.2) has a unique polynomial solution, hence

$$P_n^{(\lambda)}(x) = c \cdot P_n^{(\lambda-\frac{1}{2},\lambda-\frac{1}{2})}(x).$$

The constant c is determined by $P_n^{(\lambda)}(1) = \frac{(2\lambda)_n}{n!}$ and $P_n^{(\lambda-\frac{1}{2},\lambda-\frac{1}{2})}(1) = \binom{n+\lambda-\frac{1}{2}}{n}$. The Theorem is proved.

Theorem 4.1.3.

(4.1.13)
$$_3F_2(a,b,c;d,e;t) = \frac{\Gamma(e)t^{1-e}}{\Gamma(c)\Gamma(e-c)} \int_0^t \ {_2F_1}(a,b;d;ty)(t-y)^{e-c-1}y^{e-1}dy,$$

when $e - c > 0$.

Proof. Let $y = tY$, then $dy = tdY$, then

$$\int_0^t \ {_2F_1}(a,b;d;y)(t-y)^{e-c-1}y^{c-1}dy$$

$$= \int_0^1 \ {_2F_1}(a,b;d;tY)(t-tY)^{e-c-1}(tY)^{c-1}tdY$$

$$= t^{e-c-1+c-1+1} \int_0^1 (1-Y)^{e-c-1}Y^{c-1} \sum_{k=0}^{\infty} \frac{(a)_k(b)_k}{(d)_k}\frac{(tY)^k}{k!}dY$$

$$= t^{e-1} \sum_{k=0}^{\infty} \frac{(a)_k(b)_k}{k!(d)_k}t^k \int_0^1 (1-Y)^{e-c-1}Y^{c+k-1}dY$$

$$= t^{e-1} \sum_{k=0}^{\infty} \frac{(a)_k(b)_k}{k!(d)_k}t^k \frac{\Gamma(e-c)\Gamma(c+k)}{\Gamma(e+k)}.$$

Multiply $\frac{\Gamma(e)t^{1-e}}{\Gamma(c)\Gamma(e-c)}$ on the right hand side of the previous equation, then

$$\frac{\Gamma(e)t^{1-e}}{\Gamma(c)\Gamma(e-c)}t^{e-1} \sum_{k=0}^{\infty} \frac{\Gamma(e-c)\Gamma(c+k)}{\Gamma(e+k)} \frac{(a)_k(b)_k}{(d)_kk!}t^k$$

$$= \sum_{k=0}^{\infty} \frac{(a)_k(b)_k(c)_k}{(d)_k(e)_k}t^k = \ {_3F_2}(a,b,c;d,e;t).$$

Thus, we have proved the Theorem.

If the parameters a,b,c,d and e satisfy certain relations, then a further observation about $_3F_2$ and $_2F_1$ can be made (cf. Kazarinoff[1]).

Theorem 4.1.4 (Clausen Formula).

$$\left[{_2F_1}\left(\alpha,\beta;\alpha+\beta+\frac{1}{2};t\right)\right]^2$$

$$= {}_3F_2\left(2\alpha, 2\beta, \alpha + \beta; 2\alpha + 2\beta, \alpha + \beta + \frac{1}{2}; t\right).$$

Proof. Let $y = {}_2F_1(\alpha, \beta; \gamma; t)$ and

$$L[y] = t(1 - t)y'' + [\gamma - (\alpha + \beta + 1)t]y' - aby,$$

where $\gamma = \alpha + \beta + \frac{1}{2}$, then $L[y] \equiv 0$. Differentiating $tL[y]$, we have

$$(tL[y])' \equiv t^2(1 - t)y''' + [(\gamma + 2)t - (\alpha + \beta + 4)t^2]y''$$
$$+ [\gamma - (2\alpha + 2\beta + \alpha\beta + 2)t]y' - \alpha\beta y = 0.$$

Let $z = {}_3F_2(a, b, c; d, e; t)$ and

$$M[z] = t^2(1 - t)z''' - [(3 + a + b + c)t^2 - (1 + d + e)t]z'' +$$

(4.1.14)

$$[de - (1 + a + b + c + ab + ac + bc)t]z' - abcz,$$

then $M(z) = 0$. If $z = y^2$, then

(4.1.15) $$z' = zyy', \quad z'' = 2yy' + 2(y')^2, \quad z''' = 2yy''' + 6y'y''.$$

Replacing z, z', z'' and z''' in (4.1.14) by (4.1.15), we have $M[y^2]$. We try to find values of a, b, c, d, e and A, B such that

(4.1.16) $$M[y^2] = (2Ay + Bty')L[y] + 2y(tL[y])',$$

holds. Then $M[y^2] = 0$ provided that (4.1.16) is an identity. This implies

$$_3F_2(a, b, c; d, e; t) = c_0[{}_2F_1(\alpha, \beta; \gamma; t)]^2$$

holds, where $\gamma = \alpha + \beta + \frac{1}{2}$ and c_0 is a constant. Both sides of (4.1.16) are polynomials in y, y', y'' and y'''. These polynomials have no common factors. (4.1.16) is an identity if the coefficients of the corresponding terms in powers of y, y', y'' and y''' are equal. Thus, we have a system of equations:

$$\begin{cases} 3 + r = A + \alpha + \beta + 4, \\ 1 + v = A + \gamma + 2, \\ 1 + r + s = A(\alpha + \beta + 1) + \frac{1}{2}B\alpha\beta + 2(\alpha + \beta + 1) + \alpha\beta \\ w = \gamma(A + 1), \\ u = 2\alpha\beta(A + 1), \\ 6 = B, \\ 2(3 + r) = B(\alpha + \beta + 1), \end{cases}$$

where $r = a + b + c, s = ab + ac + bc, u = abc, \nu = d + e$ and $w = de$.

This implies

$$\begin{cases} B = 6, \\ A = 2(\alpha + \beta) - 1, \\ r = 3(\alpha + \beta), \\ \nu = 3r - 1, \\ w = (2\gamma - 1)\gamma, \\ u = 4(\alpha + \beta)\alpha\beta, \\ s = 2(\alpha + \beta)^2 + 4\alpha\beta. \end{cases}$$

Of course, a, b and c are three roots of the cubic equation:

$$x^3 - rx^2 + sx - u = 0.$$

Now, this is

$$x^3 - 3(\alpha + \beta)x^2 + (2(\alpha + \beta)^2 + 4\alpha\beta)x - 4(\alpha + \beta)\alpha\beta = 0.$$

The three roots of this equation are $2\alpha, 2\beta$ and $\alpha + \beta$.

We have $(a, b, c) = (2\alpha, 2\beta, \alpha + \beta)$.

Similarly, d and e are the roots of the quadratic equation

$$x^2 - \nu x + w = 0.$$

Now, this is

$$x^2 - (3\gamma - 1)x + (2\gamma - 1)\gamma = 0$$

The two roots of this equation are γ and $2\gamma - 1$. We have $(d, e) = (\gamma, 2\gamma - 1)$.

Finally, we have

$$_3F_2(2\alpha, 2\beta, \alpha + \beta; \gamma, 2\gamma - 1; t) = c_0 [_2F_1(\alpha, \beta; \gamma; t)]^2$$

It is easy to check $c_0 = 1$.

We have proved Theorem 4.1.4.

This theorem was established by Th. Clausen [1] in 1828.

Theorem 4.1.5 (Gegenbauer-Hua Formula).

$$P_n^\nu(x) = \sum_{k=0}^{[\frac{n}{2}]} c_k P_{n-2k}^\lambda(x)$$

where $\nu > \lambda > -\frac{1}{2}, [\frac{n}{2}]$ is the greatest integer $\leq \frac{n}{2}$, and

$$c_k(\nu, \lambda) = \frac{(n - 2k + \lambda)T(\lambda)(\nu - \lambda)_k \Gamma(n + \nu - k)}{k!\Gamma(\nu)\Gamma(n + \lambda - k + 1)}.$$

Proof. When t is sufficiently small, such that $|2xt - t^2| < 1$, we have

$$\begin{aligned}
(1 - 2xt + t^2)^{-\nu} &= \sum_{n=0}^{\infty} \frac{(\nu)_n}{n!}(2xt - t^2)^n \\
&= \sum_{n=0}^{\infty} \sum_{k=0}^{n} \frac{(-1)^k(\nu)_n(2x)^{n-k}t^{n+k}}{k!(n-k)!} \\
&= \sum_{n=0}^{\infty} t^n \sum_{k=0}^{[\frac{n}{2}]} \frac{(-1)^k(\nu)_{n-k}(2x)^{n-2k}}{k!(n-2k)!}
\end{aligned}$$

by binomial theorem. Thus

(4.1.17) $$P_n^\nu(x) = \sum_{k=0}^{[\frac{n}{2}]} \frac{(-1)^k(\nu)_{n-k}(2x)^{n-2k}}{k!(n-2k)!}.$$

Differentiating both sides of the equality

$$(1 - 2xt + t^2)^{-\lambda} = \sum_{n=0}^{\infty} P_n^\lambda(x)t^n$$

with respect to x, where $\nu > \lambda > \frac{-1}{2}$, we obtain

$$\begin{aligned}
\frac{2t\lambda}{(1 - 2xt + t^2)^{\lambda+1}} &= \sum_{n=1}^{\infty} \left(\frac{d}{dx}P_n^\lambda(x)\right)t^n \\
&= 2\lambda \sum_{m=0}^{\infty} P_m^{\lambda+1}(x)t^{m+1}.
\end{aligned}$$

Comparing the corresponding coefficients of term t^n, we have

(4.1.18) $$\frac{d}{dx}P_n^\lambda(x) = 2\lambda P_{n-1}^{\lambda+1}(x).$$

By (4.1.17), we may express $\frac{(2x)^n}{n!}$ as

(4.1.19) $$\frac{(2x)^n}{n!} = \sum_{k=0}^{[\frac{n}{2}]} a_{k,n}(\lambda)P_{n-2k}^\lambda(x).$$

Differentiating both sides of the equality

$$\frac{(2x)^{n+1}}{(n+1)!} = \sum_{k=0}^{[\frac{n+1}{2}]} a_{k,n+1}(\lambda)P_{n+1-2k}^{\lambda}(x)$$

with respect to x, we have

$$\frac{2(2x)^n}{n!} = \sum_{k=0}^{[\frac{n+1}{2}]} a_{k,n+1}(\lambda)\frac{d}{dx}P_{n+1-2k}^{\lambda}(x).$$

The left hand side of the previous equality is equal to

$$2\sum_{k=0}^{[\frac{n}{2}]} a_{k,n}(\lambda+1)P_{n-2k}^{\lambda+1}(x)$$

by (4.1.19). The right hand side of the previons equality is equal to

$$\sum_{k=0}^{[\frac{n+1}{2}]} a_{k,n+1}(\lambda)2\lambda P_{n-2k}^{\lambda+1}(x)$$

by (4.1.18). Thus we have

(4.1.20) $a_{k,n}(\lambda+1) = \lambda a_{k,n+1}(\lambda).$

   Consider

$$\frac{1}{1-2\cos\theta\cdot t+t^2} = \sum_{n=0}^{\infty} P_n^1(\cos\theta)t^n.$$

We know that

$$\frac{1}{1-2\cos\theta\cdot t+t^2} = \frac{1}{1-e^{-i\theta}t}\cdot\frac{1}{1-e^{i\theta}t} = \sum_{k=0}^{\infty} e^{-ik\theta}t^k \sum_{m=0}^{\infty} e^{im\theta}t^m$$

$$= \sum_{n=0}^{\infty}\frac{\sin(n+1)\theta}{\sin\theta}t^n.$$

By (4.1.11), $P_n^{\lambda}(x)$ is Jacobi polynoncial $P_n^{(\lambda-\frac{1}{2},\lambda-\frac{1}{2})}(x)$ multiplied by a constant. $\{P_n^{\lambda}(x)\}$ are orthogonal system with the weight function $(1-x^2)^{\lambda-\frac{1}{2}}$.

It is easy to evaluate

$$\int_{-1}^{+1} P_n^1(x)P_m^1(x)(1-x^2)^{\frac{1}{2}}dx = \int_{-\frac{\pi}{2}}^{\frac{\pi}{2}} \sin[(n+1)\theta]\sin[(m+1)\theta]d\theta = \frac{\pi}{2}\delta_{n,m}$$

and

$$\int_{-1}^{+1} \frac{(2x)^n}{n!} P_{n-2k}^1(x)(1-x^2)^{\frac{1}{2}}dx = \int_{-\frac{\pi}{2}}^{\frac{\pi}{2}} \frac{(2\cos\theta)^n}{n!}\sin[(n-2k+1)\theta]\sin\theta d\theta$$

$$= \frac{\pi}{2}\frac{n-2k+1}{k!(n-k+1)}.$$

From (4.1.19), we have

$$a_{k,n}(1) = \frac{n-2k+1}{k!(n-k+1)!}.$$

By (4.1.20), we obtain

$$a_{k,n}(\lambda) = \frac{n-2k+\lambda}{k!(\lambda)_{n-k+1}}.$$

Substituting it into (4.1.19), we have

(4.1.21)
$$\frac{(2x)^n}{n!} = \sum_{k=0}^{[\frac{n}{2}]} \frac{n-2k+\lambda}{k!(\lambda)_{n-k+1}} P_{n-2k}^\lambda(x).$$

Substituting it into (4.1.17), we obtain

$$P_n^\nu(x) = \sum_{k=0}^{[\frac{n}{2}]} c_k P_{n-2k}^\lambda(x)$$

where

$$c_k = \frac{(-1)(\nu)_{n-k}}{k!}\sum_{i=0}^{[\frac{n}{2}]-k} \frac{\lambda+n-2k-2i}{i1(\lambda)_{n-2p-i+1}}$$

$$= \sum_{p+i=k} \frac{(-1)^p(\nu)_{n-p}(\lambda+n-2p-2i)}{p!i!(\lambda)_{n-2k+i+1}}$$

or

$$\frac{\Gamma(\nu)}{\Gamma(\lambda)}c_k = \frac{\lambda+n-2k}{k!}\sum_{p=0}^{k}(-1)^p\binom{k}{p}\frac{\Gamma(\nu+n-p)}{\Gamma(\lambda+n-p+k+1)}.$$

Let $\Delta f(x) = f(x+1) - f(x)$, then (Hua [1])

$$\Delta^k\left[\frac{f(a+x)}{f(b+x)}\right] = \sum_{p=0}^{k}(-1)^p\binom{k}{p}\frac{f(a+x+k-p)}{f(b+x+k-p)}.$$

Let $x=n, a=\nu-k$ and $b=\lambda-2k+1$, we have

$$\frac{\Gamma(\nu)}{\Gamma(\lambda)}c_k = \frac{\lambda+n-2k}{k!}\Delta^k\left[\frac{\Gamma(\nu-k+n)}{\Gamma(\lambda-2k+1+n)}\right].$$

Moreover, $\Delta\left[\frac{\Gamma(a+n)}{\Gamma(b+n)}\right] = \frac{(a-b)\Gamma(a+n)}{\Gamma(b+n+1)}$ and

$$\Delta^k\left[\frac{\Gamma(a+n)}{\Gamma(b+n)}\right] = \frac{\Gamma(a-b+1)\Gamma(a+n)}{\Gamma(a-b-k+1)\Gamma(b+nk)}.$$

Thus, we obtain

$$\frac{\Gamma(\nu)}{\Gamma(\lambda)}c_k = \frac{\lambda+n-2k}{k!}\frac{\Gamma(\nu-\lambda+k)\Gamma(\nu-k+n)}{\Gamma(\nu-\lambda)\Gamma(\lambda-k+n+1)}.$$

We have proved Theorem 4.1.5.

Before we prove the important Askey-Gasper Theorem, we need to prove

Lemma 4.1.1.

(4.1.22) $$(2a)_{2j} = 2^{2j}(a)_j\left(a+\frac{1}{2}\right)_j,$$

(4.1.23) $$\sum_{k=0}^{n}\frac{(a)_k}{k!} = \frac{(a+1)_n}{n!}.$$

Proof. The formula (4.1.22) is obvious. We only need to prove (4.1.23). Obviously, it is true when $n=0$. If

$$\sum_{k=0}^{n-1}\frac{(a)_k}{k!} = \frac{(a+1)_{n-1}}{(n-1)!}$$

is true, then

$$\sum_{k=0}^{n-1} \frac{(a)_k}{k!} + \frac{(a)_n}{n!} = \frac{(a+1)_{n-1}}{(n-1)!} + \frac{(a)_n}{n!} = \frac{(a+1)_{n-1}}{(n-1)!} \left(1 + \frac{a}{n}\right) = \frac{(a+1)_n}{n!}.$$

(4.1.23) is true by induction.

Theorem 4.1.6(Askey-Gasper Theorem).

(4.1.24)
$$\sum_{k=0}^{n} P_k^{(\alpha,0)}(x) \geq 0,$$

when $\alpha > -1$ and $-1 \leq x \leq 1$.

Proof. By Theorem 4.1.1, we know

$$P_k^{(\alpha,0)}(x) = \frac{(\alpha+1)_k}{k!} \quad {}_2F_1\left(-k, k+\alpha+1; \alpha+1; \frac{1-x}{2}\right).$$

Let $a_k = \frac{(\alpha+1)_k}{k!}$, and let

$$b_{kj} = \begin{cases} \frac{(-k)_j(k+\alpha+1)_j\left(\frac{1-x}{2}\right)^j}{j!(\alpha+1)_j}, & 0 < j < k; \\ 0, & j > k. \end{cases}$$

Then

$$\sum_{k=0}^{n} P_k^{(\alpha,0)}(x) = \sum_{k=0}^{n} a_k \sum_{j=0}^{n} b_{kj} = \sum_{j=0}^{n} \sum_{k=0}^{n} a_k b_{kj}$$

$$= \sum_{j=0}^{n} \left(\sum_{k=j}^{n} a_k b_{kj}\right) = \sum_{j=0}^{n} \sum_{k=0}^{n-j} a_{k+j} b_{k+j,j}$$

$$= \sum_{j=0}^{n} \sum_{k=0}^{n-j} \frac{(\alpha+1)_{k+j}}{(k+j)!} \frac{(-k-j)_j(k+j+\alpha+1)_j}{j!(\alpha+1)_j} \left(\frac{1-x}{2}\right)^j$$

$$= \sum_{j=0}^{n} \left(\frac{x-1}{2}\right)^j \frac{(\alpha+1)_{2j}}{j!(\alpha+1)_j} \sum_{k=0}^{n-j} \frac{1}{k!} \frac{(\alpha+1)_{k+j}(k+j+\alpha+1)_j}{(\alpha+1)_{2j}}.$$

It is easy to verify that

$$\frac{(\alpha+1)_{k+j}(\alpha+1+k+j)_j}{k!(\alpha+1)_{2j}} = \frac{(\alpha+1+2j)_k}{k!}.$$

Using this equation and (4.1.23), we have

$$\sum_{k=0}^{n} P_k^{(\alpha,0)}(x) = \sum_{j=0}^{n} \left(\frac{x-1}{2}\right)^j \frac{(\alpha+1)_{2j}(\alpha+2j+2)_{n-j}}{j!(\alpha+1)_j(n-j)!}.$$

Taking $a = \frac{\alpha+1}{2}$ in (4.1.22),

$$(\alpha+1)_{2j} = 2^{2j}\left(\frac{\alpha+1}{2}\right)_j \left(\frac{\alpha+2}{2}\right)_j.$$

Substituting it into the previous equation, we obtain

$$\sum_{k=0}^{n} P_k^{(\alpha,0)}(x) = \sum_{j=0}^{n} \frac{\left(\frac{x-1}{2}\right)^j 2^{2j}\left(\frac{\alpha+1}{2}\right)_j \left(\frac{\alpha+2}{2}\right)_j (\alpha+2j+2)_{n-j}}{j!(\alpha+1)_j(n-j)!}$$

(4.1.25)
$$= \sum_{j=0}^{n} \frac{[2(x-1)]^j \left(\frac{\alpha+1}{2}\right)_j \left(\frac{\alpha+2}{2}\right)_j (\alpha+2j+2)_{n-j}}{j!(\alpha+1)_j(n-j)!}.$$

Now we need to prove

$$\sum_{k=0}^{n} P_k^{(\alpha,0)}(x) = \frac{(\alpha+2)_n}{n!}$$

(4.1.26)
$$\cdot {}_3F_2\left(-n, n+\alpha+2, \frac{\alpha+1}{2}; \alpha+1, \frac{\alpha+3}{2}; \frac{1-x}{2}\right),$$

that is, we need to prove

$$\frac{(2(x-1))^j \left(\frac{\alpha+1}{2}\right)_j \left(\frac{\alpha+2}{2}\right)_j (\alpha+2j+2)_{n-j}}{j!(\alpha+1)_j(n-j)!}$$

$$= \frac{(\alpha+2)_n}{n!} \frac{(-n)_j(n+\alpha+2)_j \left(\frac{\alpha+1}{2}\right)_j}{(\alpha+1)_j \left(\frac{\alpha+3}{2}\right)_j} \left(\frac{1-x}{2}\right)^j.$$

Equivalently, we need to prove the equation

$$\frac{1}{(n-j)!} 2^{2j}(-1)^j \left(\frac{\alpha+2}{2}\right)_j (\alpha+2j+2)_{n-j} = \frac{(\alpha+2)_n(-n)_j(n+\alpha+2)_j}{n! \left(\frac{\alpha+3}{2}\right)_j}.$$

Taking $a = \frac{\alpha+2}{2}$ in (4.1.22), we have

$$(\alpha + 2)_{2j} = 2^{2j} \left(\frac{\alpha+2}{2}\right)_j \left(\frac{\alpha+3}{2}\right)_j,$$

thus the previous equation is equivalent to

$$\frac{1}{(n-j)!}(-1)^j(\alpha+2)_{2j}(\alpha+2j+2)_{n-j} = \frac{(\alpha+2)_n(-n)_j(n+\alpha+2)_j}{n!}.$$

Obviously, the equalities $(a)_{2j}(a+2j)_{n-j} = (a)_n(a+n)_j$ and $(-1)^j(-n)_j = \frac{n!}{(n-j)!}$ are true. Using these equalities for $a = \alpha+2$, we can verify that the previous equation is true. This proves (4.1.26).

Let $c = \frac{\alpha+1}{2}$ and $e = 2c = \alpha+1$ in Theorem 4.1.3. Then

$$\frac{\Gamma(e)t^{1-e}}{\Gamma(c)\Gamma(e-c)} = \frac{\Gamma(2c)t^{1-e}}{(\Gamma(c))^2} = \frac{\Gamma(\alpha+1)t^{-\alpha}}{\left[\Gamma\left(\frac{\alpha+1}{2}\right)\right]^2},$$

$e - c = \frac{\alpha+1}{2}$, and (4.1.13) becomes

$$_3F_2\left(a, b, \frac{\alpha+1}{2}; d, \alpha+1; t\right)$$

$$= t^{-\alpha}\Gamma(\alpha+1)\left[\Gamma\left(\frac{\alpha+1}{2}\right)\right]^{-2}\int_0^t [(t-y)y]^{\frac{\alpha-1}{2}} \; _2F_1(a, b; d; y)dy.$$

The integral exists since $\alpha > -1$. If we define the linear operator L as

$$Lg = t^{-\alpha}\Gamma(\alpha+1)\left[\Gamma\left(\frac{\alpha+1}{2}\right)\right]^{-2}\int_0^t [(t-y)y]^{\frac{\alpha-1}{2}} g(y)dy, \quad \alpha > -1, \; t > 0,$$

then

(4.1.27) $$_3F_2\left(a, b, \frac{\alpha+1}{2}; d, \alpha+1; t\right) = L(_2F_1(a, b; d; t)).$$

Let $2\lambda = \alpha+2$ and $x = 1 - 2t$ in Theorem 4.1.2, it becomes
(4.1.28)
$$P_n^{(\lambda)}(x) = P_n^{\left(\frac{\alpha+2}{2}\right)}(1-2t) = \frac{(\alpha+2)_n}{n!}F\left(-n, n+\alpha+2; \frac{\alpha+3}{2}; t\right).$$

Let $\mu = \frac{\alpha+2}{2}$ and $\lambda = \frac{\alpha+1}{2}$ in Theorem 4.1.5. (Gegenbauer-Hua Theorem), then

$$(4.1.29) \quad P_n^{\left(\frac{\alpha+2}{2}\right)}(1-2t) = \sum_{j=0}^{\left[\frac{n}{2}\right]} \frac{\left(\frac{\alpha}{2}\right)_j \left(\frac{\alpha+2}{2}\right)_{n-j} \left(\frac{\alpha+3}{2}\right)_{n-2j}}{j! \left(\frac{\alpha+3}{2}\right)_{n-j} \left(\frac{\alpha+1}{2}\right)_{n-2j}} P_{n-2j}^{\left(\frac{\alpha+1}{2}\right)}(1-2t).$$

holds. From Theorem 4.1.1, we know
$$(4.1.30)$$
$$P_{n-2j}^{\left(\frac{\alpha+1}{2}\right)}(1-2t) = \frac{(\alpha+1)_{n-2j}}{(n-2j)!} F\left(-n+2j, n-2j+\alpha+1; \frac{\alpha+2}{2}; t\right).$$

Combining (4.1.28), (4.1.29) and (4.1.30), we obtain

$$\frac{(\alpha+2)_n}{n!} F\left(-n, n+\alpha+2; \frac{\alpha+3}{2}; t\right)$$

$$= \sum_{j=0}^{\left[\frac{n}{2}\right]} \frac{\left(\frac{\alpha}{2}\right)_j \left(\frac{\alpha+2}{2}\right)_{n-j} \left(\frac{\alpha+3}{2}\right)_{n-2j} (\alpha+1)_{n-2j}}{j! \left(\frac{\alpha+3}{2}\right)_{n-j} \left(\frac{\alpha+1}{2}\right)_{n-2j} (n-2j)!}$$

$$(4.1.31) \qquad \cdot F\left(-n+2j, n-2j+\alpha+1; \frac{\alpha+2}{2}; t\right).$$

By the definition of L and (4.1.27), we have

$${}_3F_2\left(-n, n+\alpha+2, \frac{\alpha+1}{2}; \frac{\alpha+3}{2}, \alpha+1; t\right) =$$

$$L\left[F\left(-n, n+\alpha+2; \frac{\alpha+3}{2}; t\right)\right],$$

and

$${}_3F_2\left(-n+2j, n-2j+\alpha+1, \frac{\alpha+1}{2}; \frac{\alpha+2}{2}, \alpha+1; t\right)$$

$$= L\left[F\left(-n+2j, n-2j+\alpha+1; \frac{\alpha+2}{2}; t\right)\right].$$

By (4.1.26), (4.1.31) and the previous two formulas, it follows that

$$\sum_{k=0}^{n} P_k^{(\alpha,0)}(x) = \frac{(\alpha+2)_n}{n!} L\left[F\left(-n, n+\alpha+2; \frac{\alpha+3}{2}; t\right)\right]$$

$$= \sum_{j=0}^{\left[\frac{n}{2}\right]} \frac{\left(\frac{\alpha}{2}\right)_j \left(\frac{\alpha+2}{2}\right)_{n-j} \left(\frac{\alpha+3}{2}\right)_{n-2j} (\alpha+1)_{n-2j}}{j! \left(\frac{\alpha+3}{2}\right)_{n-j} \left(\frac{\alpha+1}{2}\right)_{n-2j} (n-2j)!}$$

$$(4.1.32) \qquad \cdot {}_3F_2\left(-n+2j, n-2j+\alpha+1, \frac{\alpha+1}{2}; \frac{\alpha+2}{2}, \alpha+1; t\right),$$

because L is a linear operator. Let $2a = -n + 2j$ and $2b = n - 2j + \alpha + 1$ in Theorem 4.1.4.(Clausen formula):

$$\left[F\left(a, b; a + b + \frac{1}{2}; t \right) \right]^2 = {}_3F_2 \left[2a, 2b, a + b; 2a + 2b, a + b + \frac{1}{2}; t \right],$$

it becomes

$$\left[F\left(\frac{-n + 2j}{2}, \frac{n - 2j + \alpha + 1}{2}; \frac{\alpha + 2}{2}; t \right) \right]^2$$

$$= {}_3F_2 \left(-n + 2j, n - 2j + \alpha + 1, \frac{\alpha + 1}{2}; \alpha + 1, \frac{\alpha + 2}{2}; t \right).$$

Since $\alpha > -1$, the coefficients of each term in (4.1.32) are positive, which implies that

$$\sum_{k=0}^{n} P_k^{(\alpha, 0)}(x) \geq 0.$$

§4.2. De Branges Theorem

Now we consider the resolution of the Bieberbach Conjecture. We will prove the famous de Branges Theorem; that is we will give de Branges' proof of the Milin conjecture. As mentioned at the end of Chapter one, a proof of the Milin conjecture implies the Robertson conjecture which implies the Bieberbach Conjecture.

A *special function system of de Branges* is introduced as follows: Let $n = 1, 2, \cdots$, for a fixed n, we define:

$$\tau_{n,k}(t) = k \sum_{\nu=0}^{n-k} (-1)^\nu \frac{(2k + \nu + 1)_\nu (2k + 2\nu + 2)_{n-k-\nu}}{(k + \nu)\nu!(n - k - \nu)!} e^{-(\nu + k)t},$$

(4.2.1) $$\tau_{n,n+1}(t) \equiv 0.$$

where $k = 1, 2, \ldots, n,$

Lemma 4.2.1. *If $P_j^{(\alpha,\beta)}(x)$ are Jacobi polynomials, then*

$$(4.2.2) \qquad \tau'_{n,k}(t) = -ke^{-kt}\sum_{j=0}^{n-k} P_j^{(2k,0)}(1-2e^{-t}).$$

Proof. By definition (4.2.1), we have

$$(4.2.3) \qquad \frac{-\tau'_{n,k}(t)e^{kt}}{k} = \sum_{\nu=0}^{n-k}(-1)^\nu \frac{(2k+\nu+1)_\nu(2k+\nu+2)_{n-k-\nu}}{\nu!(n-k-\nu)!}e^{-\nu t}.$$

Taking $\alpha = 2k, x = 1 - 2e^{-t}$ in (4.1.32), and replacing n by $n-k$, we find

$$\sum_{\nu=0}^{n-k} P_\nu^{(2k,0)}(1-2e^{-t})$$

$$= \sum_{\nu=0}^{n-k}(-1)^\nu \frac{2^{2\nu}e^{-\nu t}\left(\frac{2k+1}{2}\right)_\nu\left(\frac{2k+2}{2}\right)_\nu(2k+2\nu+2)_{n-k-\nu}}{(2k+1)_\nu \nu!(n-k-\nu)!}.$$

We know that $2^{2\nu}\left(\frac{2k+1}{2}\right)_\nu\left(\frac{2k+2}{2}\right)_\nu = (2k+1)_{2\nu}$ and

$$\frac{(2k+1)_{2\nu}}{(2k+1)_\nu} = (2k+\nu+1)_\nu$$

by (4.1.22). Thus, the previous equation becomes

$$\sum_{\nu=0}^{n-k} P_\nu^{(2k,0)}(1-2e^{-t}) = \sum_{\nu=0}^{n-k}(-1)^\nu \frac{(2k+\nu+1)_\nu(2k+2\nu+2)_{n-k-\nu}}{\nu!(n-k-\nu)!}e^{-\nu t}.$$

Comparing with (4.2.3), we obtain (4.2.2).

As a consequence of Lemma 4.2.1. and the Askey-Gasper Theorem (4.1.24), we have

Theorem 4.2.1. *If $0 \le t \le \infty$, and $k = 1, 2, \cdots$, then*

$$(4.2.4) \qquad\qquad\qquad \tau'_{n,k}(t) \le 0$$

holds, for $k = 1, 2, \cdots, n$.

Moreover, we have

Theorem 4.2.2. *The special function system of de Branges satisfies the following relations:*

$$(4.2.5) \qquad \tau_{n,k}(t) - \tau_{n,k+1}(t) = -\frac{\tau'_{n,k}(t)}{k} - \frac{\tau'_{n,k+1}(t)}{k+1}, \quad k = 1, \cdots, n,$$

$$(4.2.6) \qquad \tau_{n,k}(0) = n - k + 1, \quad k = 1, 2, \cdots, n,$$

$$\tau_{n,n+1}(t) \equiv 0. \quad 0 \le t < \infty$$

Proof. For each $n = 1, 2, \cdots$, since $\tau_{n,n+1} \equiv 0$, it is easy to see that the differential equations (4.2.5) with initial conditions (4.2.6) define a unique system. Solve for $\tau_{n,n}(t)$, then for $\tau_{n,n-1}$, $\cdots$.

Note that de Branges' system does require that $\tau_{n,n+1} \equiv 0$ and satisfies the initial condition (4.2.6). To show that (4.2.5) is satisfied, it is equivalent to showing that $\tau_{n,k} + \frac{\tau'_{n,k}}{k} = \tau_{n,k+1} - \frac{\tau'_{n,k+1}}{k+1}$. Let

$$(4.2.7) \qquad G_k = \frac{\tau_{n,k} e^{kt}}{k}, \quad V_k = \frac{\tau_{n,k} e^{-kt}}{k}, \quad 1 \le k \le n+1.$$

Since $G'_k = \left(\frac{\tau'_{n,k}}{k} + \tau_{n,k}\right) e^{kt}$ and $V'_k = \left(\frac{\tau'_{n,k}}{k} - \tau_{n,k}\right) e^{-kt}$, it suffices to show that

$$(4.2.8) \qquad G'_k e^{-kt} = -V'_{k+1} e^{(k+1)t}, \quad 1 \le k \le n.$$

By (4.2.1) and (4.2.7), we have

$$G_k = \sum_{\nu=0}^{n-k} (-1)^\nu \frac{(2k+\nu+1)_\nu (2k+2\nu+2)_{n-k-\nu}}{(k+\nu)\nu!(n-k-\nu)!} e^{-\nu t},$$

$$V_k = \sum_{\nu=0}^{n-k} (-1)^\nu \frac{(2k+\nu+1)_\nu (2k+2\nu+2)_{n-k-\nu}}{(k+\nu)\nu!(n-k-\nu)!} e^{-\nu t - 2kt}.$$

We can get (4.2.8) by evaluating G'_k and V'_{k+1} directly from these two formulas. This proves (4.2.5).

It is easy to prove that $P_n^{(\alpha,\beta)}(-1) = (-1)^n P_n^{(\beta,\alpha)}(1)$, and hence $P_j^{(\alpha,0)}(-1) = (-1)^j$. We have $\tau'_k(0) = -k \sum_{j=0}^{n-k} (-1)^j$ by Lemma 4.2.1. Then

$$(4.2.9) \qquad -\frac{\tau'_{n,k}(0)}{k} = \sum_{j=0}^{n-k} (-1)^j = \begin{cases} 1, & when \quad n-k \quad is \quad even; \\ 0, & when \quad n-k \quad is \quad odd. \end{cases}$$

By (4.2.5) and (4.2.9), we have $\tau_{n,k}(0) - \tau_{n,k+1}(0) = 1,\ k = 1, 2, \cdots, n$. Equation (4.2.6) follows if we add the previous formulas.

Now we can prove the famous de Branges Theorem. (de Branges [1], [2], [3], FitzGerald and Pommerenke[1])

Theorem 4.2.3(de Branges Theorem). *The Milin Conjecture (1.3.13) is true, with the equality holding if and only if $f(z)$ is the Koebe function or one of its rotations.*

Proof. Without loss of generality, we need only consider a function $f(z) \in S$ such that $f(z)$ maps $|z| < 1$ onto the complex plane with a slit, and the slit is a Jordan curve tending to infinity. As we mentioned in Chapter 2, such functions are dense in S. We need only prove this theorem for such functions. By the results of section 2.2, we know that for such function, there exists a chain of functions

$$f(z, t) = e^t z + \cdots + a_n(t) z^n + \cdots, \quad |z| < 1, \quad 0 \le t < \infty,$$

such that $f(z, t)$ satisfies the Löwner differential equation

(2.2.10)
$$\frac{\partial f(z, t)}{\partial t} = \frac{1 + \kappa(t) z}{1 - \kappa(t) z} z \frac{\partial f(z, t)}{\partial z},$$

and $f(z, 0) = f(z)$, where $|\kappa(t)| = 1$, $\kappa(t)$ is a continuous function on $0 \le t < \infty$.

Let

(4.2.10)
$$\log\left(\frac{f(z, t)}{e^t z}\right) = \sum_{k=1}^{\infty} c_k(t) z^k, \quad |z| < 1,$$

when $0 \le t < \infty$, then $c_k(0) = 2\gamma_k$, where γ_k is defined by (1.3.8).

Differentiating (4.2.10) with respect to t and z respectively, we have

(4.2.11)
$$\frac{f_t(z, t)}{f(z, t)} - 1 = \sum_{k=1}^{\infty} c'_k(t) z^k,$$

(4.2.12)
$$\frac{f_z(z, t)}{f(z, t)} - \frac{1}{z} = \sum_{k=1}^{\infty} k c_k(t) z^{k-1}.$$

From (2.2.10), (4.2.11) and (4.2.12), we have

$$1 + \sum_{k=1}^{\infty} c_k'(t)z^k = \frac{1 + \kappa(t)z}{1 - \kappa(t)z}\left(1 + \sum_{k=1}^{\infty} kc_k(t)z^k\right)$$

$$= (1 + 2\kappa(t)z + 2\kappa(t)^2z^2 + \cdots)\left(1 + \sum_{k=1}^{\infty} kc_k(t)z^k\right).$$

Comparing the corresponding coefficients, we obtain

$$c_k'(t) = 2\kappa(t)^k + kc_k(t) + 2\sum_{j=1}^{k-1} \kappa(t)^{k-j}jc_j(t).$$

Let

$$(4.2.13) \qquad b_0(t) \equiv 0, \quad b_k(t) = \sum_{j=1}^{k} jc_j(t)\kappa(t)^{-j}, \quad k = 1, 2, \cdots,$$

then

$$(4.2.14) \qquad c_k'(t) = 2\kappa(t)^k - kc_k(t) + 2\kappa(t)^k b_k(t), \quad k = 1, 2, \cdots.$$

For a fixed n, define

$$(4.2.15) \qquad \varphi(t) = \sum_{k=1}^{n}\left(k|c_k(t)|^2 - \frac{4}{k}\right)\tau_{n,k}(t),$$

then the Milin conjecture is

$$(4.2.16) \qquad \varphi(0) - \sum_{k=1}^{n}\left(k|c_k(0)|^2 - \frac{4}{k}\right)(n - k + 1) \leq 0$$

by (4.2.6). The idea of the proof of (4.2.16) is as follows: we need only prove

$$(4.2.17) \qquad \varphi'(t) = -\sum_{k=1}^{n}|b_{k-1}(t) + b_k(t) + 2|^2 \frac{\tau_{n,k}'(t)}{k}.$$

for $t \geq 0$.

If (4.2.17) is true, then $\varphi'(t) \geq 0$ by Theorem 4.2.1, and hence $\varphi(t)$ is a monotonic increasing function. By the compactness of S and the definition

(4.2.1) of $\{\tau_{n,k}(t)\}$, we have $\varphi(\infty) = 0$. This implies that $\varphi(0) \leq 0$ which is exactly the Milin Conjecture.

We will make some simple algebric calculations and use (4.2.14) and Theorem 4.2.2 to prove (4.2.17) from (4.2.15).

We know $\kappa(t)^{-1} = \bar{\kappa}(t)$ since $|\kappa(t)| = 1$, and

$$(4.2.18) \qquad \begin{cases} \bar{b}_k(t) - \bar{b}_{k-1}(t) = k\bar{c}_k(t)\kappa(t)^k, \\ b_k - b_{k-1} = kc_k\overline{\kappa(t)^k}, \\ |b_k - b_{k-1}|^2 = k^2|c_k|^2, \end{cases}$$

by (4.2.13). From (4.2.15) and (4.2.18), we have

$$(4.2.19) \qquad \varphi(t) = \sum_{k=1}^{n}(|b_k - b_{k-1}|^2 - 4)\frac{\tau_{n,k}(t)}{k}.$$

Let $kc_k(t) = \kappa^k(b_k(t) - b_{k-1}(t)) = u(t)$, then $|u|^2 = |b_k - b_{k-1}|^2$ and

$$\frac{\partial |u|^2}{\partial t} = u_t\bar{u} + u\bar{u}_t = 2\Re e(u_t\bar{u}) = 2\Re e(kc_k'(t)\bar{u})$$
$$= 2\Re e[kc_k'(t)\kappa(t)^{-k}(\bar{b}_k(t) - \bar{b}_{k-1}(t))],$$

that is,

$$(4.2.20) \qquad \frac{\partial}{\partial t}\left(\frac{|b_k - b_{k-1}|^2}{k}\right) = 2\Re e\{c_k'(t)\kappa(t)^{-k}(\bar{b}_k - \bar{b}_{k-1})\}.$$

Substituting (4.2.14) into (4.2.20), we have

$$\frac{\partial}{\partial t}\left(\frac{|b_k - b_{k-1}|^2}{k}\right) = 2\Re e\{2(\bar{b}_k - \bar{b}_{k-1}) - kc_k\kappa^{-k}(\bar{b}_k - \bar{b}_{k-1}) + 2b_k(\bar{b}_k - \bar{b}_{k-1})\}.$$

We know $kc_k\kappa^{-k} = b_k - b_{k-1}$ by (4.2.18), hence

$$\frac{\partial}{\partial t}\left(\frac{|b_k - b_{k-1}|^2}{k}\right) = 2\Re e\{2(\bar{b}_k - \bar{b}_{k-1}) - |b_k - b_{k-1}|^2 + 2b_k(\bar{b}_k - \bar{b}_{k-1})\}$$
$$= -2|b_k - b_{k-1}|^2 + 4\Re e\{(1 + b_k)(\bar{b}_k - \bar{b}_{k-1})\}.$$

Differentiating (4.2.19) and using the previous equation, we have

$$\varphi'(t) = \sum_{k=1}^{n}(|b_k - b_{k-1}|^2 - 4)\frac{\tau_{n,k}'(t)}{k}$$
$$+ \sum_{k=1}^{n}\tau_{n,k}(t)[-2|b_k - b_{k-1}|^2 + 4\Re e(1 + b_k)(\bar{b}_k - \bar{b}_{k-1})].$$

Since

$$-\sum_{k=1}^{n}(2|b_k|^2 + 4\Re eb_k)\tau_{n,k+1} = -\sum_{k=2}^{n+1}(2|b_{k-1}|^2 + 4\Re eb_{k-1})\tau_{n,k}$$

$$= -\sum_{k=1}^{n}(2|b_{k-1}|^2 + 4\Re eb_{k-1})\tau_{n,k},$$

we obtain

$$\sum_{k=1}^{n}(2|b_k|^2 + 4\Re eb_k)(\tau_{n,k} - \tau_{n,k+1})$$

$$= \sum_{k=1}^{n}(2|b_k|^2 + 4\Re eb_k - 2|b_{k-1}|^2 - 4\Re eb_{k-1})\tau_{n,k}.$$

Obviously,

$$-2|b_k - b_{k-1}|^2 + 4\Re e(1+b_k)(\bar{b}_k - \bar{b}_{k-1}) = 2|b_k|^2 - 2|b_{k-1}|^2 + 4\Re eb_k - 4\Re eb_{k-1}.$$

This proves that

$$\varphi'(t) = \sum_{k=1}^{n}(|b_k - b_{k-1}|^2 - 4)\frac{\tau'_{n,k}}{k} + \sum_{k=1}^{n}(2|b_k|^2 + 4\Re eb_k)(\tau_{n,k} - \tau_{n,k+1})$$

since $\tau_{n,n+1}^{(t)} = 0$ and $b_0(t) = 0$. From (4.2.5), we have

$$\varphi'(t) = \sum_{k=1}^{n}(|b_k - b_{k-1}|^2 - 4)\frac{\tau'_{n,k}}{4} + \sum_{k=1}^{n}(2|b_k|^2 + 4\Re eb_k)\left(-\frac{\tau'_{n,k}}{k} - \frac{\tau'_{n,k+1}}{k+1}\right)$$

$$= \sum_{k=1}^{n}[(|b_k - b_{k-1}|^2 - 4) - 2|b_k|^2 - 4\Re eb_k]\frac{\tau'_{n,k}}{k}$$

$$- \sum_{k=2}^{n+1}(2|b_{k-1}|^2 + 4\Re eb_{k-1})\frac{\tau'_{n,k}}{k}$$

$$= \sum_{k=1}^{n}\frac{\tau'_k}{k}[|b_k - b_{k-1}|^2 - 4 - 2|b_k|^2 - 4\Re eb_k - 2|b_{k-1}|^2 - 4\Re eb_{k-1}]$$

$$= -\sum_{k=1}^{n}|b_k + b_{k-1} + 2|^2\frac{\tau'_{n,k}}{k}$$

since $\tau_{n,n+1}(t) = 0$ and $b_0(t) = 0$.

As noted before, this proves the Milin Conjecture.

Now we will prove that equality holds in the Milin Conjecture if and only if $f(z)$ is the Koebe function or one of its rotations.

If $f(z) \in S$, but is not the Koebe function or one of its rotations, then $|a_2| < 2$ by Theorem 1.1.3. We choose a sequence of functions $f_m \in S$, such that f_m maps $|z| < 1$ onto the complex plane minus a Jordan curve which tends to infinity, and the sequence f_m is uniformly convergent to f on any compact subset of $|z| < 1$. For f_m, we have the corresponding functions $f_m(z,t)$. The coefficients $a_{n,m}(t)$ of the expansion of $f_m(z,t)$ correspond to the coefficients $c_{n,m}(t)$ at (4.2.10). When m is sufficiently large, there exists α such that

$$(4.2.21) \qquad |c_{1,m}(0)| = |a_{2,m}(0)| < \alpha < 2,$$

From (4.2.14), we have

$$|c'_{1,m}(t)| = |c_{1,m}(t) + 2\kappa(t)| \leq |a_{2,m}(t)| + 2 \leq 4.$$

Thus $|c_{1,m}(t)| \leq \alpha + 4t$ for $t \geq 0$. Note $\alpha + 4t \leq 2$ for $0 \leq t \leq \frac{2-\alpha}{4}$. By (4.2.21) and $\tau'_k(t) < 0$, we have

$$\varphi'_m(t) \geq |c_{1,m}(t)\bar{k}_m(t) + 2|^2(-\tau'_1(t)) \geq (2 - \alpha - 4t)^2(-\tau'_1(t))$$

when $0 \leq t \leq \frac{2-\alpha}{4}$, and m is sufficiently large. Since

$$-\int_0^\infty \varphi'_m(t)dt = \varphi_m(0) = \sum_{k=1}^n \left(k|c_{k,m}(0)|^2 - \frac{4}{k} \right)(n+1-k),$$

we have

$$\sum_{k=1}^n \left(k|c_{k,m}|^2 - \frac{4}{k} \right)(n+1-k)$$

$$\leq -\int_0^{\frac{2-\alpha}{8}} \varphi'_m(t)dt \leq \left(\frac{2-\alpha}{2} \right)^2 \int_0^{\frac{2-\alpha}{8}} \tau'_1(t)dt$$

$$= \left(\frac{2-\alpha}{2} \right)^2 \left[\tau_1\left(\frac{2-\alpha}{8} \right) - \tau_1(0) \right] < 0.$$

Let $m \to \infty$, then

$$\sum_{k=1}^n \left(k|c_k|^2 - \frac{4}{k} \right)(n+1-k) < 0.$$

Thus equality holds in the Milin Conjecture if and only if $f(z)$ is the Koebe function or one of its rotations. This proves the de Branges Theorem.

The de Branges Theorem implies the Robertson Conjecture and hence the Bieberbach Conjecture. Actually, de Branges proved a result more general than the proof of the Milin Conjecture. He proved that: If $f(z) \in S$, and $|f(z)| < e^T$, then

$$\sum_{k=1}^{n} k(n+1-k)|c_k + q_k|^2 \leq \sum_{k=1}^{n} k|p_k|^2 \tau_{n,k}(T) + \sum_{k=1}^{n} \frac{4}{k}(n+1-k-\tau_{n,k}(T))$$

holds, where p_k is any complex number, and q_k is defined by

$$\sum_{k=1}^{\infty} q_k z^k = \sum_{k=1}^{\infty} p_k e^{-kT} f(z)^k.$$

When $p_k = 0$ and $T \to \infty$, this is the Milin Conjecture.

The proof of this result is similar to the proof of the Milin Conjecture. We can get this result by considering the coefficients of the expansion of the function

$$\log \frac{f(z,t)}{e^t z} + \sum_{k=1}^{\infty} p_k e^{-kT} f(z,t)^k,$$

defining the corresponding function φ, and then integrating $-\varphi'$ from 0 to T.

From the de Branges Theorem, we have the following consequence:

Theorem 4.2.4(Landau Theorem). *If $f \in S$, then*

(4.2.22) $$|f^{(n)}(z)| \leq K^{(n)}(|z|) = n! \frac{n+|z|}{(1-|z|)^{n+2}},$$

where $K(z)$ is the Koebe function, and equality holds if and only if f is the Koebe function or one of its rotations.

In fact, Landau (E. Landau [2]) proved that the Bieberbach Conjecture and (4.2.22) were equivalent in 1925. It is obvious that (4.2.22) implies the Bieberbach Conjecture. We need only prove the Bieberbach Conjecture implies (4.2.22).

If $f \in S$, $0 \le r < 1$, $F(z) = f\left(\frac{z+r}{1+rz}\right) = \sum_{\nu=0}^{\infty} c_\nu z^\nu$, then

$$f(z) = F\left(\frac{z-r}{1-rz}\right) = \sum_{\nu=0}^{\infty} c_\nu \left(\frac{z-r}{1-rz}\right)^\nu$$

and

$$\frac{f^{(n)}(r)}{n!} = \sum_{\nu=1}^{n} \frac{c_n}{(n-\nu)!}\frac{d^{n-\nu}}{dz^{n-\nu}}(1-rz)^{-\nu}|_{z=r} = \sum_{\nu=1}^{n} c_\nu P_\nu(r,n),$$

where $P_\nu(r,n)$ is non-negative and independent of f. Because

$$\frac{F(z) - F(0)}{F'(0)} \in S,$$

we have

$$|c_\nu| \le \nu|F'(0)| = \nu(1-r^2)|f(r)| \le \nu\left(\frac{1+r}{1-r}\right)^2$$

when $2 \le \nu \le n$. Thus

$$\frac{|f^{(n)}(r)|}{n!} \le \sum_{\nu=1}^{n} \nu\left(\frac{1+r}{1-r}\right)^2 P_\nu(r,n).$$

Let $K(z)$ be the Koebe function $\frac{z}{(1-z)^2}$. Then the coefficients c_ν of the expansion of $K\left(\frac{z+r}{1+rz}\right)$ are $\nu\left(\frac{1+r}{1-r}\right)^2$. Therefore

$$\frac{|f^{(n)}(r)|}{n!} \le \sum_{\nu=1}^{n} \nu\left(\frac{1+r}{1-r}\right)^2 P_\nu(r,n) = \frac{K^{(n)}(r)}{n!} = \frac{n+r}{(1-r)^{n+2}}.$$

We obtain (4.2.22) if we replace f by $e^{-i\alpha}f(e^{i\alpha}z)$.

The Poincaré-Bergman metric of the unit disk is $ds^2 = \frac{2|dz|^2}{(1-|z|^2)^2}$ (cf. §1.1), its connection is $\Gamma = \frac{2\bar{z}}{1-|z|^2}$, and the corresponding covariant derivative is ∇. As a consequence of de Branges Theorem, we have

Theorem 4.2.5. *If $f(z) \in S$, then*

(4.2.23) $$\left|\frac{\nabla^n f(z)}{\nabla f(z)}\right| \le \left|\frac{\nabla^n K(|z|)}{\nabla K(|z|)}\right| = \frac{n \cdot n!}{(1-|z|^2)^{n-1}},$$

where $K(z)$ is the Koebe function, and equality holds if and only if $f(z)$ is the Koebe function or one of its rotations.

This can be written (4.2.23) in the form of intrinsic derivatives as

(4.2.24)
$$\left| \frac{\delta^n f}{\delta s^n} \right| \leq n! n \left| \frac{\delta f}{\delta s} \right|.$$

The intrinsic derivative is an invariant, thus (4.2.24) is an estimate for an invariant. It is the geometric meaning of the de Branges Theorem.

In order to prove Theorem 4.2.5, we need only prove the following three Lemmas:

Lemma 4.2.2. *If $f(z) = \sum_{n=0}^{\infty} a_n z^n$ is holomorphic at $|z| < 1$, then*

(4.2.25)
$$f\left(\frac{\zeta + z}{1 + \bar{z}\zeta} \right) = \sum_{n=0}^{\infty} g_n(z) \zeta^n,$$

where $|\zeta| < 1$, $g_0(z) = f(z)$ and

(4.2.26)
$$g_n(z) = \sum_{j=0}^{n-1} \frac{(-1)^j c_{n-1}^j}{(n-j)!} \bar{z}^j (1 - |z|^2)^{n-j} f^{(n-j)}(z)$$

when $n \geq 1$.

Proof. At first, we prove the following equation by induction.

$$\frac{\partial^n}{\partial \zeta^n} f\left(\frac{\zeta + z}{1 + \bar{z}\zeta} \right) = f^{(n)}\left(\frac{\zeta + z}{1 + \bar{z}\zeta} \right) \left(\frac{1 - |z|^2}{(1 + \bar{z}\zeta)^2} \right)^n$$

(4.2.27)
$$+ \sum_{j=1}^{n-1} A_j^{(n)} \frac{\bar{z}^j (1 - |z|^2)^{n-j}}{(1 + \bar{z}\zeta)^{2n-j}} f^{(n-j)}\left(\frac{z + \zeta}{1 + \bar{z}\zeta} \right),$$

where $A_j^{(n)} = (-1)^j \frac{n(n-1)^2 \cdots (n-j+1)^2 (n-j)}{j!}$.

Obviously,

$$\frac{\partial}{\partial \zeta} f\left(\frac{\zeta + z}{1 + \bar{z}\zeta} \right) = f'\left(\frac{\zeta + z}{1 + \bar{z}\zeta} \right) \frac{1 - |z|^2}{(1 + \bar{z}\zeta)^2}$$

holds true, i.e., (4.2.27) is true when $n = 1$. If (4.2.27) is true for $n \leq k$, then

$$\frac{\partial^{k+1}}{\partial \zeta^{k+1}} f \left(\frac{\zeta + z}{1 + \bar{z}\zeta} \right)$$

$$= \frac{\partial}{\partial \zeta} \left[f^{(k)} \left(\frac{\zeta + z}{1 + \bar{z}\zeta} \right) \left(\frac{1 - |z|^2}{(1 + \bar{z}\zeta)^2} \right)^k \right.$$

$$\left. + \sum_{j=1}^{k-1} A_j^{(k)} \frac{\bar{z}^j (1 - |z|^2)^{k-j}}{(1 + \bar{z}\zeta)^{2k-j}} f^{(k-j)} \left(\frac{\zeta + z}{1 + \bar{z}\zeta} \right) \right]$$

$$= f^{(k+1)} \left(\frac{\zeta + z}{1 + \bar{z}\zeta} \right) \left(\frac{1 - |z|^2}{(1 + \bar{z}\zeta)^2} \right)^{k+1}$$

$$+ f^{(k)} \left(\frac{\zeta + z}{1 + \bar{z}\zeta} \right) \frac{(1 - |z|^2)^k}{(1 + \bar{z}\zeta)^{2k+1}} (-2k\bar{z})$$

$$+ A_1^{(k)} \frac{\bar{z}(1 - |z|^2)^k}{(1 + \bar{z}\zeta)^{(2k+1)}} f^{(k)} \left(\frac{\zeta + z}{1 + \bar{z}\zeta} \right)$$

$$+ \sum_{j=2}^{k-1} [A_j^{(k)} - (2k - j + 1)A_{j-1}^{(k)}]\bar{z}^j \frac{(1 - |z|^2)^{k-j+1}}{(1 + \bar{z}\zeta)^{2(k+1)-j}} f^{(k+1-j)} \left(\frac{\zeta + z}{1 + \bar{z}\zeta} \right)$$

$$+ A_{k-1}^{(k)} \bar{z}^k (-k - 1) \frac{1 - |z|^2}{(1 + \bar{z}\zeta)^{k+2}} f' \left(\frac{\zeta + z}{1 + \bar{z}\zeta} \right).$$

We can prove

$$\begin{cases} A_1^{(k+1)} = A_1^{(k)} - 2k, \\ A_j^{(k+1)} = A_j^{(k)} - (2k - j + 1)A_{j-1}^{(k)}, \quad (2 \leq j \leq k - 1) \\ A_k^{(k+1)} = A_{k-1}^{(k)}(-k - 1), \end{cases}$$

by direct calculation. Using these relations in the previous equation, we have (4.2.27) for the case $n = k + 1$.

Letting $\zeta = 0$ in (4.2.27), we have (4.2.26).

Lemma 4.2.3. *If $f(z)$ is holomorphic at $|z| < 1$, then*

$$(4.2.28) \qquad \nabla^n f(z) = \frac{n! g_n(z)}{(1 - |z|^2)^n},$$

is true for any positive integers, when $g_n(z)$ is defined by (4.2.26).

Proof. We prove (4.2.28) by induction.

Equation (4.2.28) is true when $n = 1$, since $\nabla f = f'$. If (4.2.28) is true for $n \leq k$, then

$$\nabla^{k+1} f = \nabla(\nabla^k f) = \frac{d}{dz}(\nabla^k f) - k\Gamma\nabla^k f = \frac{d}{dz}(k! g_k(z)(1 - |z|^2)^{-k})$$
$$- k\Gamma k! g_k(z)(1 - |z|^2)^{-k} = k!(1 - |z|^2)^{-k-1}((1 - |z|^2)g_k' - k\bar{z}g_k).$$

Differentiating both sides of (4.2.26), we get

$$g_k' = \left[\sum_{j=0}^{k-1} \frac{(-1)^j c_{k-1}^j}{(k-j)!} \bar{z}^j (1 - |z|^2)^{k-j} f^{(k-j)}(z)\right]'$$
$$= \sum_{j=0}^{k-1} \frac{(-1)^j c_{k-1}^j}{(k-j)!} \bar{z}^j [(k-j)(1 - |z|^2)^{k-j-1}(-\bar{z}) f^{(k-j)}(z)$$
$$+ (1 - |z|^2)^{k-j} f^{(k+1-j)}(z)]$$
$$= \sum_{j=0}^{k-1} \frac{(-1)^{j+1} c_{k-1}^j}{(k-1-j)!} \bar{z}^{j+1} (1 - |z|^2)^{k-1-j} f^{(k-j)}(z)$$
$$+ \sum_{j=0}^{k-1} \frac{(-1)^j c_{k-1}^j}{(k-j)!} \bar{z}^j (1 - |z|^2)^{k-j} f^{(k+1-j)}(z).$$

Thus $\nabla^{k+1} f$ is equal to

$$\frac{k!}{1 - |z|^2} \sum_{j=0}^{k-1} (-1)^{j+1} \left(\frac{c_k^{j+1}}{(k-1-j)!} + \frac{k c_{k-1}^j}{(k-j)!}\right)$$
$$\times \bar{z}^{j+1}(1 - |z|^2)^{k-j} f^{(k-j)}(z)$$
$$+ \frac{k!}{1 - |z|^2} \left[(k+1)(-1)^k \bar{z}^k (1 - |z|^2) f'(z)\right.$$
$$\left. + \frac{(1 - |z|^2)^{k+1}}{k!} f^{(k+1)}(z)\right].$$

We have

$$\nabla^{k+1} f = \frac{(k+1)!}{(1 - |z|^2)^{k+1}} g_{k+1}(z)$$

by the following identity

$$\frac{c_k^{j+1}}{(k-1-j)!} + \frac{kc_{k-1}^j}{(k-j)!} = \frac{(k+1)c_k^{j+1}}{(k-j)!}.$$

This proves (4.2.28).

By Lemma 4.2.3, Lemma 4.2.2. and the definition of the intrinsic derivative, we have

Lemma 4.2.4. *If* $f(z) = \sum_{n=0}^{\infty} a_n z^n$ *is holomorphic at* $|z| < 1$, *then*

(4.2.29) $$f\left(\frac{\zeta+z}{1+\bar{z}\zeta}\right) = \sum_{n=0}^{\infty} \frac{(1-|z|^2)^n \nabla^n f(z)}{n!} \zeta^n,$$

and

(4.2.30) $$f\left(\frac{\zeta+z}{1+\bar{z}\zeta}\right) = \sum_{n=0}^{\infty} \frac{1}{n!} \frac{\delta^n f(z)}{\delta s^n} \zeta^n$$

where $|\zeta| < 1$.

Using the de Branges Theorem on (4.2.29) and (4.2.30), we have (4.2.23) and (4.2.24) in Theorem 4.2.5.

From (4.2.23) and (4.2.24), we may obtain many interesting results (cf. Gong [3], Gong and Yan [1]). Here we state one of them: If $f \in S, 0 \leq \alpha \leq \beta \leq 2\pi, 0 \leq r < 1$, then the inequality

$$\left| \frac{\delta^n f(re^{i\alpha})}{\delta s^n} - \frac{\delta^n f(re^{i\beta})}{\delta s^n} \right| \leq (n+1)!(n+1)L(f;r;\alpha,\beta)$$

holds, where $L(f;r;\alpha,\beta)$ is the length of the $arc(re^{i\alpha}, re^{i\beta})$ on the circle $|z| = r$ under the mapping $w = f(z)$.

§4.3. Weinstein's Proof

After de Branges' proof of the Milin conjecture in 1984, there were many efforts to understand this important development. Survey papers tried to put this breakthrough into the context of previous work and to explain the proof. There were efforts to simplify the proof of the key special function result of Askey and Gasper.

In 1991, Weinstein(L. Weinstein[1]) gave a short, elegent proof of the de Branges theorem. The proof is based on properties of Legendre polynomials, instead of the Jacobi polynomials which had been used by Askey and Gasper.

In order to state Weinstein's proof, we need to state and to prove some classical results about Legendre polynomials. For a more detailed discussion of Legendre polynomials, see e.g., Whittaker and Watson (E. T. Whittaker and G. N. Watson[1]).

Let x be a real number, z be a complex number and $|2xz - z^2| < 1$, we can expand $(1 - 2xz + z^2)^{-\frac{1}{2}}$ into a series of ascending powers of $2xz - z^2$, and then expand it into a series of powers of z,

$$(4.3.1) \quad (1 - 2xz + z^2)^{-\frac{1}{2}} = P_0(x) + zP_1(x) + z^2 P_2(x) + z^3 P_3(x) + \cdots$$

where $P_0(x) = 1$, $P_1(x) = x$, $P_2(x) = \frac{1}{2}(3x^2 - 1)$, $P_3(x) = \frac{1}{2}(5x^3 - 3x)$, $\cdots$, and generally

$$P_n(x) = \sum_{r=0}^{[\frac{n}{2}]} (-1)^r \frac{(2n - 2r)!}{2^n r!(n - r)!(n - 2r)!} x^{n-2r}.$$

The expressions $P_0(x)$, $P_1(x)$, $\cdots$, which are polynomials in x, are known as Legendre polynomials. $P_n(x)$ is called the *Legendre polynomial of degree* n. (Compare with Gegenbaur p dynomial defined by (4.1.8))

It is evident that, when n is an integer,

$$\frac{d^n}{dx^n}(x^2 - 1)^n = \sum_{r=0}^{[\frac{n}{2}]} (-1)^r \frac{n!}{r!(n - r)!} \frac{(2n - 2r)!}{(n - 2r)!} x^{n-2r}.$$

From (4.3.1), it follows that

$$(4.3.2) \qquad P_n(x) = \frac{1}{2^n n!} \frac{d^n}{dx^n}(x^2 - 1)^n,$$

and this result is known as Rodrigues' formula.

Using the integral representation of the n-th derivative of a holomorphic function, it follows that

$$(4.3.3) \qquad P_n(x) = \frac{1}{2\pi i} \int_C \frac{(t^2 - 1)^n}{2^n (t - x)^{n+1}} dt,$$

where C is a contour which encircles the point x once counter- clockwise. This is the Schläfli formula for Legendre polynomials.

We can express the Legendre polynomials as hypergeometric functions. Suppose $|1 - x| \leq 2(1 - \delta)$, $0 < \delta < 1$, and we take C to be the circle $|1 - t| = 2 - \delta$, since

$$\left| \frac{1 - x}{1 - t} \right| \leq \frac{2 - 2\delta}{2 - \delta} < 1,$$

we may expand $(t - x)^{-n-1}$ in the uniformly convergent series

$$(t - x)^{-n-1} = (t - 1)^{-n-1}$$
$$\times \left\{ 1 + (n + 1)\frac{x - 1}{t - 1} + \frac{(n + 1)(n + 2)}{2!} \left(\frac{x - 1}{t - 1} \right)^2 + \cdots \right\}.$$

Substituting this result in the Schläfli integral (4.3.3), and integrating term-by-term, we have

$$P_n(x) = \sum_{r=0}^{\infty} \frac{(x - 1)^r}{2^{n+1}\pi i} \frac{(n + 1)(n + 2) \cdot (n + r)}{r!} \int_C \frac{(t^2 - 1)^n}{(t - 1)^{n+1+r}} dt$$

$$= \sum_{r=0}^{\infty} \frac{(x - 1)^r (n + 1)(n + 2) \cdots (n + r)}{2^n (r!)^2} \left[\frac{d^r}{dt^r} (t + 1)^n \right]_{t=1}.$$

Since

$$\left[\frac{d^r}{dt^r} (t + 1)^n \right]_{t=1} = 2^{n-r} n(n - 1) \cdot (n - r + 1),$$

we have

$$P_n(x) = \sum_{r=0}^{\infty} \frac{(n + 1)(n + 2) \cdots (n + r)(-n)(1 - n) \cdots (r - 1 - n)}{(r!)^2}$$

(4.3.4)

$$\times (\frac{1}{2} - \frac{1}{2}x)^r = \quad {}_2F_1(n + 1, -n; 1; \frac{1}{2} - \frac{1}{2}x)$$

when $|1 - x| \leq 2(1 - \delta) < 2$. From (4.3.4), we immediately get

(4.3.5) $$P_n(x) = P_{-n-1}(x).$$

Let k be a positive integer and $-1 < x < 1$, n being unrestricted, the function

(4.3.6) $$P_n^k(x) = (1 - x^2)^{\frac{k}{2}} \frac{d^k P_n(x)}{dx^k}$$

will be called the *Ferrer associated Legendre function of degree n and order k.* By (4.3.3) and (4.3.6), we have

(4.3.7)
$$P_n^k(x) = \frac{(n+1)(n+2)\cdots(n+k)}{2^{n+1}\pi i}(1-x^2)^{\frac{k}{2}}\int_C (t^2-1)^n(t-x)^{-n-k-1}dt.$$

Let $t = x + (x^2-1)^{\frac{1}{2}}e^{i\phi}$, i.e. we take C as a circle centered at x with radius $|(x^2-1)^{\frac{1}{2}}|$, then $dt = (x^2-1)^{\frac{1}{2}}e^{i\phi}id\phi$, $t^2-1 = 2(x^2-1)^{\frac{1}{2}}e^{i\phi}((x^2-1)\cos\phi+x)$, $t-x = (x^2-1)^{\frac{1}{2}}e^{i\phi}$. Substituting these results into (4.3.7), it is

$$P_n^k(x) = \frac{(n+1)(n+2)\cdots(n+k)}{2\pi}(-1)^{\frac{k}{2}}$$

$$\times \int_{-\pi}^{\pi}[(x^2-1)^{\frac{1}{2}}\cos\phi+x]^n e^{-ik\phi}d\phi.$$

But $(x^2-1)\cos\phi+x$ is an even function of ϕ which implies

$$\int_{-\pi}^{\pi}[(x^2-1)\cos\phi+x]^n\sin k\phi d\phi = 0.$$

Therefore,

$$P_n^k(x) = \frac{(n+1)(n+2)\cdots(n+k)}{2\pi}(-1)^{\frac{k}{2}}$$

(4.3.8)
$$\times \int_{-\pi}^{\pi}[(x^2-1)^{\frac{1}{2}}\cos\phi+x]^n\cos k\phi d\phi.$$

We will show the addition theorem for Legendre polynomials

(4.3.9)
$$P_n(x) = P_n(u)P_n(v) + 2\sum_{k=1}^{n}\frac{(n-k)!}{(n+k)!}P_n^k(u)P_n^k(v)\cos k\theta,$$

where $x = uv - (u^2-1)^{\frac{1}{2}}(v^2-1)^{\frac{1}{2}}\cos\theta$.

For a fixed v, if $v > 0$, it is easy to prove that

$$\left|\frac{u+(u^2-1)^{\frac{1}{2}}\cos(\theta-\varphi)}{v+(v^2-1)^{\frac{1}{2}}\cos\varphi}\right|$$

is a bounded function of φ. Let M be its upper bound and $|z| < M^{-1}$, then

$$\sum_{n=0}^{\infty}z^n\frac{[u+(u^2-1)^{\frac{1}{2}}\cos(\theta-\varphi)]^n}{[v+(v^2-1)^{\frac{1}{2}}\cos\varphi]^{n+1}}$$

converges uniformly with respect to φ, and thus

$$\sum_{n=0}^{\infty} z^n \int_{-\pi}^{\pi} \frac{[u + (u^2 - 1)^{\frac{1}{2}} \cos(\theta - \varphi)]^n}{[v + (v^2 - 1)^{\frac{1}{2}} \cos\varphi]^{n+1}} d\varphi$$

$$= \int_{-\pi}^{\pi} \sum_{n=0}^{\infty} z^n \frac{[u + (u^2 - 1)^{\frac{1}{2}} \cos(\theta - \varphi)]^n}{v + (v^2 - 1)^{\frac{1}{2}} \cos\varphi]^{n+1}} d\varphi$$

$$= \int_{-\pi}^{\pi} [v - zu + ((v^2 - 1)^{\frac{1}{2}} - z(u^2 - 1)^{\frac{1}{2}} \cos\theta) \cos\varphi$$

$$(4.3.10) \qquad - z(u^2 - 1)^{\frac{1}{2}} \sin\theta \sin\varphi]^{-1} d\varphi.$$

It is also easy to verify that

$$\int_{-\pi}^{\pi} \frac{d\varphi}{A + B\cos\varphi + C\sin\varphi} = \frac{2\pi}{(A^2 - B^2 - C^2)^{\frac{1}{2}}}$$

where the value of the radical is taken which makes

$$|A - (A^2 - B^2 - C^2)^{\frac{1}{2}}| < |(B^2 + C^2)^{\frac{1}{2}}|.$$

Thus, the integral on the right hand side of (4.3.10) is equal to

$$2\pi[(v - zu)^2 - [(v^2 - 1)^{\frac{1}{2}} - z(u^2 - 1)^{\frac{1}{2}} \cos\theta]^2 - [z(u^2 - 1)^{\frac{1}{2}} \sin\theta]^2]^{-\frac{1}{2}}$$

$$(4.3.10') \qquad = \frac{2\pi}{(1 - 2zx + z^2)^{\frac{1}{2}}}.$$

Comparing with the definition of the Legendre polynomials, we have

$$(4.3.11) \qquad P_n(x) = \frac{1}{2\pi} \int_{-\pi}^{\pi} \frac{[u + (u^2 - 1)^{\frac{1}{2}} \cos(\theta - \varphi)]^n}{[v + (v^2 - 1)^{\frac{1}{2}} \cos\varphi]^{n+1}} d\varphi.$$

Since $P_n(x)$ is a polynomial of degree n in $\cos\theta$, we can expand it as a Fourier cosine series

$$P_n(x) = \frac{1}{2}A_0 + \sum_{k=1}^{n} A_k \cos k\theta$$

where

$$A_k = \frac{1}{\pi} \int_{-\pi}^{\pi} P_n(x) \cos k\theta \, d\theta$$

$$= \frac{1}{2\pi^2} \int_{-\pi}^{\pi} \int_{-\pi}^{\pi} \frac{[u + (u^2 - 1)^{\frac{1}{2}} \cos(\theta - \varphi)]^n \cos k\theta}{[v + (v^2 - 1)^{\frac{1}{2}} \cos\varphi]^{n+1}} d\theta d\varphi$$

$$= \frac{1}{2\pi^2} \int_{-\pi}^{\pi} \int_{-\pi}^{\pi} \frac{[u + (u^2 - 1)^{\frac{1}{2}} \cos\psi]^n \cos k(\varphi + \psi)}{[v + (v^2 - 1)^{\frac{1}{2}} \cos\varphi]^{n+1}} d\psi d\varphi$$

$$= \frac{1}{2\pi^2} \int_{-\pi}^{\pi} \int_{-\pi}^{\pi} \frac{[u + (u^2 - 1)^{\frac{1}{2}} \cos\psi]^n \cos k\varphi \cos k\psi}{[v + (v^2 - 1)^{\frac{1}{2}} \cos\varphi]^{n+1}} d\psi d\varphi$$

since

$$\int_{-\pi}^{\pi} [u + (u^2 - 1)^{\frac{1}{2}} \cos \psi]^n \sin k\psi d\psi = 0.$$

By (4.3.5) and (4.3.8), we have

$$A_k = 2 \frac{(n-k)!}{(n+k)!} P_n^k(u) P_n^k(v).$$

We have proved (4.3.9).

Now we can state the proof of the de Branges Theorem by Weinstein.

Fix $z \in D$; define $w = w_t(z)$ by $\frac{z}{(1-z)^2} = \frac{e^t w}{(1-w)^2}$, $t \geq 0$. We have

$$\sum_{n=1}^{\infty} \left(\sum_{k=1}^{n} \left(\frac{4}{k} - k|c_k(0)|^2 \right) (n-k+1) \right) z^{n+1}$$

$$= \sum_{k=1}^{\infty} \sum_{n=k}^{\infty} \left(\frac{4}{k} - k|c_k(0)|^2 \right) (n-k+1) z^{n+1}$$

$$= \sum_{k=1}^{\infty} \sum_{m=1}^{\infty} m z^{m+k} \left(\frac{4}{k} - k|c_k(0)|^2 \right)$$

$$= \frac{z}{(1-z)^2} \sum_{k=1}^{\infty} \left(\frac{4}{k} - k|c_k(0)|^2 \right) z^k$$

$$= -\int_0^{\infty} -\frac{z}{(1-z)^2} \frac{d}{dt} \left(\sum_{k=1}^{\infty} \left(\frac{4}{k} - k|c_k(t)|^2 \right) w^k \right) dt$$

$$= \int_0^{\infty} \frac{e^t w}{1-w^2} \frac{1+w}{1-w} \left(\sum_{k=1}^{\infty} k(c_k(t)\bar{c}_k(t))' w^k \right.$$

(4.3.12)

$$\left. + \sum_{k=1}^{\infty} (4 - k^2|c_k(t)|^2) w^k \frac{1-w}{1+w} \right) dt,$$

since $w_t(z) \to 0$ when $t \to \infty$, and $\frac{\partial w}{\partial t} = -w\frac{1-w}{1+w}$.

By (4.2.11), we have

$$c_k'(t) = \lim_{r \to 1} \frac{1}{2\pi} \int_0^{2\pi} \frac{\frac{\partial f(z_1, t)}{\partial t}}{f(z_1, t)} \bar{z}_1^k d\theta.$$

where $z_1 = re^{i\theta}$. The right hand side of (4.3.12) equals

$$
\int_0^\infty \frac{e^t w}{1 - w^2} \left[\frac{1+w}{1-w} \left(1 + \sum_{k=1}^\infty \left(\lim_{r \to 1} \frac{1}{2\pi} \int_0^{2\pi} \frac{\frac{\partial f(z_1,t)}{\partial t}}{f(z_1,t)} k\bar{c}_k(t)\bar{z}_1^k d\theta \right) w^k \right) \right.
$$

$$
+ \frac{1+w}{1-w} \left(1 + \sum_{k=1}^\infty \left(\lim_{r \to 1} \frac{1}{2\pi} \int_0^{2\pi} \frac{\frac{\partial \bar{f}(z_1,t)}{\partial t}}{\bar{f}(z_1,t)} kc_k(t)z_1^k d\theta \right) w^k \right)
$$

$$
\left. - 2 \left(\frac{1+w}{1-w} \right) + \frac{4w}{1-w} - \sum_{k=1}^\infty k^2 |c_k(t)|^2 w^k \right] dt = \int_0^\infty \frac{e^t w}{1 - w^2} \left[1 \right.
$$

$$
+ \sum_{k=1}^\infty \left(\lim_{r \to 1} \frac{1}{2\pi} \int_0^{2\pi} \frac{\frac{\partial f(z_1,t)}{\partial t}}{f(z_1,t)} (2(1 + \cdots + k\bar{c}_k(t)\bar{z}_1^k) - k\bar{c}_k(t)\bar{z}_1^k) d\theta \right)
$$

$$
\times w^k + 1 + \sum_{k=1}^\infty \left(\lim_{r \to 1} \frac{1}{2\pi} \int_0^{2\pi} \frac{\frac{\partial \bar{f}(z_1,t)}{\partial t}}{\bar{f}(z_1,t)} (2(1 + \cdots + kc_k(t)z_1^k) \right.
$$

(4.3.13)

$$
\left. \left. - kc_k(t)z_1^k) d\theta \right) w^k - 2 - \sum_{k=1}^\infty k^2 |c_k(t)|^2 w^k \right] dt.
$$

By (4.2.12), the right hand side of (4.3.13) equals

$$
\int_0^\infty \frac{e^t w}{1 - w^2} \left(\sum_{k=1}^\infty \lim_{r \to 1} \frac{1}{2\pi} \int_0^{2\pi} \left\{ \frac{\frac{\partial f(z_1,t)}{\partial t}}{f(z_1,t)} \frac{f(z_1,t)}{z_1 \frac{\partial f(z_1,t)}{\partial z_1}} \right\} \left(1 + \sum_{l=1}^\infty lc_l(t)z_1^l \right) \right.
$$

$$
\times (2(1 + \cdots + k\bar{c}_k(t)\bar{z}_1^k) - k\bar{c}_k(t)\bar{z}_1^k) d\theta w^k
$$

$$
+ \sum_{k=1}^\infty \lim_{r \to 1} \frac{1}{2\pi} \int_0^{2\pi} \left\{ \frac{\frac{\partial \bar{f}(z_1,t)}{\partial t}}{\bar{f}(z_1,t)} \frac{\bar{f}(z_1,t)}{\bar{z}_1 \frac{\partial \bar{f}(z_1,t)}{\partial \bar{z}_1}} \right\} \left(1 + \sum_{l=1}^\infty l\bar{c}_l(t)\bar{z}_1^l \right)
$$

$$
\times (2(1 + \cdots + kc_k(t)z_1^k) - kc_k(t)z_1^k) d\theta w^k
$$

(4.3.14)

$$
\left. - \sum_{k=1}^\infty k^2 |c_k(t)|^2 w^k \right) dt.
$$

By the Löwner differential equation (2.2.10), we have

$$
\frac{\partial f(z,t)}{\partial t} \left(z \frac{\partial f(z,t)}{\partial z} \right)^{-1} = \frac{1 + \kappa(t)z}{1 - \kappa(t)z}
$$

and

$$(1 + \sum_{l=1}^{k} lc_l(t)z_1^l)(2(1 + \cdots + k\bar{c}_k(t)\bar{z}_1^k) - k\bar{c}_k(t)\bar{z}_1^k)$$

$$= \frac{1}{2}|2(1 + \cdots + kc_k(t)z_1^k) - kc_k(t)z_1^k|^2 + kc_k(t)z_1^k(1 + \cdots$$

$$+ (k-1)\bar{c}_{k-1}(t)\bar{z}_1^{k-1}) + \frac{1}{2}k^2|c_k(t)|^2r^{2k}.$$

The expression (4.3.14) equals

$$\int_0^\infty \frac{e^t w}{1 - w^2} \sum_{k=1}^\infty \lim_{r \to 1} \frac{1}{2\pi} \int_0^{2\pi} \Re\left\{\frac{1 + \kappa(t)z_1}{1 - \kappa(t)z_1}\right\} |2(1 + \cdots + kc_k(t)z_1^k)$$

(4.3.15)

$$- kc_k(t)z_1^k|^2 d\theta w^k dt = \int_0^\infty \frac{e^t w}{1 - w^2} \left(\sum_{k=1}^\infty A_k(t)w^k\right) dt,$$

where

$$A_k(t) = \lim_{r \to 1} \frac{1}{2\pi} \int_0^{2\pi} \Re\left\{\frac{1 + \kappa(t)z_1}{1 - \kappa(t)z_1}\right\} |2(1 + \cdots + kc_k(t)z_1^k)$$

(4.3.15')

$$- kc_k(t)z_1^k|^2 d\theta \geq 0$$

for $t \geq 0$, $k = 1, 2, \cdots$, since $\Re\left\{\frac{1+\kappa(t)z_1}{1-\kappa(t)z_1}\right\} \geq 0$.

Let

(4.3.16)
$$\frac{e^t w^{k+1}}{1 - w^2} = \sum_{n=0}^\infty \Lambda_k^n(t)z^{n+1},$$

then by (4.3.15), we will complete the proof of the Milin conjecture if we can prove $\Lambda_k^n(t) \geq 0$ for $t \geq 0$.

Let $u = v = (1 - e^{-t})^{\frac{1}{2}}$ in (4.3.9), then $x = 1 - e^{-t} + e^{-t}\cos\theta$ and

$$P_n^R(u) = P_n^k(v) = e^{-\frac{tk}{2}}\left(\frac{d}{dt}\right)^n P_n(t)$$

by (4.3.6). (4.3.9) yields

$$P_n(x) = \sum_{k=0}^\infty A_{k,n}(t)\cos k\theta$$

where $A_{k,n}(t) \geq 0$. By (4.3.1), we have

$$\frac{z}{1 - 2xz + z^2} = z \left\{ \sum_{n=0}^{\infty} P_n(x) z^n \right\}^2$$

$$= z \left\{ \sum_{m=0}^{\infty} z^m \sum_{k=0}^{m} A_{k,m} \cos k\theta \right\} \cdot \left\{ \sum_{n=0}^{\infty} z^n \sum_{l=0}^{n} A_{l,n} \cos n\theta \right\}.$$

Using the elementary formula $cos k\theta cos l\theta = \frac{1}{2} \left\{ \cos(k - \ell)\theta + \cos(k + \ell)\theta \right\}$ at the right side of the preveons equality, we obtain

$$(4.3.17) \qquad \frac{z}{1 - 2xz + z^2} = \sum_{n=0}^{\infty} z^{n+1} \sum_{k=0}^{n} B_{k,n} \cos k\theta$$

where $B_{k,n} \geq 0$.

We recall that $w = w_t(z)$ was defined by

$$\frac{z}{(1 - z)^2} = \frac{e^t w}{(1 - w)^2}.$$

Thus

$$\frac{z}{1 - 2xz + z^2} = \frac{1}{1 + \frac{1}{z} - 2x} = \frac{1}{2 + e^{-t} \left(\frac{1}{w} + w - 2 \right) - 2x}$$

$$= \frac{e^t w}{1 - 2w \cos \theta + w^2} = \frac{e^t w}{1 - w^2} \cdot \frac{1 - w^2}{1 + w^2 - 2w \cos \theta}$$

$$= \frac{e^t w}{1 - w^2} \left\{ 1 + 2 \sum_{k=1}^{\infty} w^k \cos k\theta \right\}$$

$$(4.3.18) \qquad = \sum_{k=0}^{\infty} \Lambda_0^n(t) z^{n+1} + 2 \sum_{k=1}^{\infty} \sum_{n=k}^{\infty} \Lambda_k^n(t) z^{n+1} \cos k\theta.$$

Equating the coefficients of z^{n+1} and $z^{n+1} \, cos k\theta$ in (4.1.17) and (4.1.18) respectively, we obtain

$$\Lambda_0^n = B_{0,n}, \quad \text{and} \quad \Lambda_k^n = \frac{1}{2} B_{k,n}, \quad k \geq 1.$$

We have proved $\Lambda_k^n(t) \geq 0$ for $t \geq 0, k = 0, 1, \cdots, n$.

Weinstein's proof as stated in this section is same as in his paper except the last part, the proof of $\Lambda_k^n(t) \geq 0$, which was given by Hayman (cf.

Hayman [1] Chap. 8). His proof may be influenced by a short paper by Exhad and Zilberger [1].

The case of equality for the Milin Conjecture follows from considering $A_1(t)$ and using the fact that $|a_2| = 2$ implies that the function is the Koebe function or one of its rotations.

Thus the proof of Weinstein is complete.

Having presented Weinstein's proof, we will present a related work of Wilf. In 1994, he wrote a paper with the title " A footnote on two proofs of the Bieberbach-De Branges theorem". Wilf says: *The functions that Weinstein encountered, but did not identify, are none other than the same polynomials of Askey and Gasper that de Branges had met. In particular, then, Weinstein's argument gives an independent proof of the nonnegativity of Askey-Gasper polynomials.* (H. S. Wilf[1] p.61).

Now we are going to recount this "footnote" of Wilf which shows the close connection between Weinstein's Λ_k^n and the polynomials that Askey-Gasper studied.

We know

$$(4.3.19) \qquad G^{(1)}(x, w) = \frac{1}{1 - 2xw + w^2} = \sum_{n=0}^{\infty} P_n^{(1)}(x) w^n,$$

by (4.1.12), where $-1 \le x \le 1$ and $|w| < 1$. By Theorem 4.1.2, (4.1.11) and the definition of hypergeometric function (4.1.4), we have

$$P_n^{(1)}(x) = {}_2F_1(-n, n + 2; \frac{3}{2}, \frac{1 - x}{2}) \frac{(2)_n}{n!}$$

$$= \sum_{r=0}^{\infty} \frac{(-n)_r (n + 2)_r (1 - x)^r (2)_n}{\left(\frac{3}{2}\right)_r 2^r r! n!}$$

when $|x| < 1$ and $n \ge 1$. It is easy to verify

$$\frac{(-n)_r (n + 2)_r (2)_n}{\left(\frac{3}{2}\right)_r 2^2 n!} = \frac{(n + r + 1)!(-1)^r}{(n - r)!(r + \frac{1}{2})(r - \frac{1}{2}) \cdots (\frac{3}{2}) 2^r r!}$$

$$= \frac{(n + r + 1)!(-1)^r 2^r}{(n - r)!(2r + 1)!} = \binom{n + r + 1}{2r + 1} (-1)^r 2^r.$$

We have

$$(4.3.20) \qquad P_n^{(1)}(x) - \sum_r \binom{n + r + 1}{2r + 1} 2^r (x - 1)^r.$$

By the definition of $w_t(z)$, it is easy to verify the equation

$$\frac{z}{1 - 2(\cos^2\phi + \sin^2\phi\cos\theta)tz^2} = \frac{e^t w}{1 - w^2} + 2\sum_{k=1}^{\infty}\frac{e^t w^{k+1}}{1 - w^2}\cos k\theta$$

holds where $\sin\phi = e^{-\frac{t}{2}}$.

Let $c = \cos^2\phi$, then

$$\frac{1}{1 - 2z(c + (1-c)\cos\theta) + z^2} = \sum_{n=0}^{\infty}\sum_{k=0}^{n}\Lambda_k^n(c)z^n\cos k\theta,$$

by (4.3.16), and we find

$$(4.3.21) \qquad P_n^{(1)}(c + (1-c)\cos\theta) = \sum_{k=0}^{n}\Lambda_k^n(c)\cos k\theta$$

by (4.3.19). Differentiating (4.3.21) r times with respect to c and put $c = 1$, we have

$$(1 - \cos\theta)^r(P_n^{(1)})^{(r)}(1) = \sum_{m=0}^{n}(\Lambda_m^n)^{(r)}(1)\cos m\theta.$$

If we multiply by $\cos k\theta$ and integrate from 0 to 2π, we obtain

$$\pi(\Lambda_k^n)^{(r)}(1) = 2^{r+1}(P_n^{(1)})^{(r)}(1)\int_0^{\pi}(\sin\theta)^{2r}\cos 2k\theta d\theta.$$

For integer $r \geq 0$ and $k \geq 0$, it is easy to check

$$\int_0^{\pi}(\sin t)^{2r}\cos 2kt dt = \frac{(-1)^k}{4^r}\binom{2r}{r+k},$$

hence

$$(\Lambda_k^n)^{(r)}(1) = \frac{(-1)^k(2r)!(P_n^{(1)})^{(r)}(1)}{2^{r-1}(r-k)!(r+k)!} = \frac{(-1)^k}{2^{r-1}}\binom{2r}{r+k}(P_n^{(1)})^{(r)}(1).$$

We obtain

$$\Lambda_k^n(c) = 2(-1)^k\sum_{r=0}^{n}\binom{2r}{r+k}\binom{n+r+1}{2r+1}(c-1)^r$$

$$= 2(-1)^k\sum_{r=0}^{n}\frac{(n+r+1)!(c-1)^r}{(r+k)!(r-k)!(n-r)!(2r+1)}.$$

If we let $r = k + s$, then

$$\Lambda_k^n(c) = 2(-1)^k \sum_{s=0}^{n-k} \frac{(n+k+s+1)!(c-1)^{k+s}}{(2k+1)!s!(n-k-s)!(2k+2s+1)}$$

$$= (1-c)^k \sum_{s=0}^{n-k} \frac{(n+k+2)_s(k-n)_n(k+\frac{1}{2})_s}{(2k+1)_s(k+\frac{3}{2})_s s!}$$

$$\times \frac{\Gamma(n+k+2)(1-c)^s}{\Gamma(n-k+1)\Gamma(2k+1)(k+\frac{1}{2})}$$

$$= 2(1-c)^k \sum_{s=0}^{n-k} \frac{(n+k+2)_s(k-n)_s(k+\frac{1}{2})_s}{(2k+1)_s(k+\frac{3}{2})_s} \frac{(1-c)^s}{s!} \binom{n+k+1}{2k+1}$$

$$= 2(1-c)^k \binom{n+k+1}{2k+1}$$

(4.3.22)
$$\times \quad {}_3F_2(-n+k, n+k+2, k+\frac{1}{2}; 2k+1, k+\frac{3}{2}; 1-c).$$

These are the polynomials which Askey and Gasper studied. Weinstein proved that ${}_3F_2$ at (4.3.22) is nonnegative. Actually, he proved

$$\Lambda_k^n(\cos^2 \phi) = 2 \sum_{l+m=n} \frac{(m-k)!}{(m+k)!}(P_l(\cos\phi))^2(P_m^k(\cos\phi))^2$$

$$+ 2\left\{ \sum_{h+j=k, l+m=n} + \sum_{h-j=k, l+m=n} \right\} \frac{(l-h)!(m-j)!}{(l+h)!(m+j)!}$$

$$\times (P_l^h(\cos\phi))^2(P_m^j(\cos\phi))^2 \geq 0,$$

in his paper (Weinstein [1]).

In 1993, Todorov (P. G. Todorov [1]) pointed out more connections between these two proofs. He proved that $\Lambda_k^n(t) = \frac{-\tau_{n,k}'(t)}{k}$, where $\Lambda_k^n(t)$ is defined by (4.3.16) and $\tau_{n,k}(t)$ is defined by (4.2.1). He also proved that

(4.3.23) $$\sum_{k=1}^{n} \left(\frac{4}{k} - k|c_k(0)|^2\right)(n-k+1) = \int_0^{\infty} \sum_{k=1}^{n} A_k(t)\frac{\tau_{nk}'}{k} dt$$

where $A_k(t)$ is defined by (4.3.15'). Of course, any one of these two equalities implies Milin conjecture is true again due to $\tau_{n,k}'(t) < 0$ and

$A_k(t) \geq 0$. (4.3.23) is easy to obtain from FitzGerald and Pommerenke's paper (C. H. FitzGerald and Ch. Pommerenke [1]).

After de Branges' proof, many papers were written to make the significance of his proof more understandable from various different points of view. For example, Koornwinder (T. H. Koornwinder [1]) has researched the group-theoretic interpretations of this proof. He offers a less computational and more conceptual proof. Nikolskii and Vasyunin (N. K. Kikolskii and V. I. Vasyunin [1],[2]) give a functional analysis explanation of the equations (4.2.5). de Branges (L. de Branges [7]) and Helton and Weening (J. H. Helton and F. Weening [1]) develop the proof in systems views. Rovneyak (J. Rornyak [1]) makes some connections of the proof with Krein space operator theory, etc.

After de Branges' proof, many papers have been devoted to a step-by-step simplification of the original de Branges' proof. Until now, no one has been able to avoid the Milin conjecture to prove the Bieberbach conjecture or the Robertson conjecture directly.

After de Branges' proof, very many papers have used his ideas, methods and results to get new results. It is difficult to list these papers exhaustively.

Many textbooks or monographs include the details of the proof of de Branges Theorem. For example, Henrici's book (P. Henrici [1]) in 1986, Hayman's book (W. Hayman [1]) in 1994, Rosenblum and Rovnyak's book (M. Rosenblum and J. Rovnyak [1]) in 1994, and Conway's book (J. B. Conway [1]) in 1995, etc.

CHAPTER V

SEVERAL COMPLEX VARIABLE CASES

§5.1. Counter-example

Geometrical function theory of one complex variable has a long history and a large number of important and interesting results. But there are counter-examples which show that many of these results are not true in several complex variables.

Perhaps H. Cartan was the first mathematician to systematically extend geometrical function theory from one variable to several variables. In 1933, in P. Montel's book on univalent function theory, Henri Cartan [1] wrote an appendix entitled "Sur la possibilite d'entendre aux fonctions de plusieurs variables complexes la theorie des fonctions univalents" in which he called for a number of generalizations of properties of univalent functions in one variable to biholomorphic mappings in several complex variables. He pointed out that **there does not exist a corresponding Bieberbach conjecture in the case of several complex varialbles** even in the simplest situation and that the boundedness of the modulus of the second coefficient of the Taylor expansions of the normalized univalent functions on the unit disc is not true in several complex variables. He also demonstrated that the corresponding growth and covering theorems fail in the case of several complex variables.

We exhibit the following counter-example.

For any positive integer k, let $f(z) = (f_1(z), f_2(z)), z = (z_1, z_2) \in \mathbb{C}^2$, with
$$\begin{cases} f_1(z) = z_1, \\ f_2(z) = z_2(1 - z_2)^{-k} = z_2 + kz_1z_2 + \cdots . \end{cases}$$

Then f is a normalized biholomorphic mapping on the unit ball $B^2 = \{z \in \mathbb{C}^2 : z\bar{z}' < 1\}$ ($\bar{z}'$ means the conjugate and transposition of z) in $\mathbb{C}^2$, that is,

$f(0) = 0$, the Jacobian J_f of f at $z = 0$ is identity matrix.

It is easy to observe the modulus of the coefficents of the second order terms in the Taylor expansion of $f(z)$, the growth $|f(z)|$, and the distortion of $f(z)$, $|det J_f(z)|$, are unbounded.

The above example is a very simple counter-example. Usually, there are many coefficients of the same order terms in the Taylor expansion of a biholomorphic mapping of several complex variables. We may ask the following question. One coefficient of a term in the Taylor expansion is unbounded. Can we make a combination of coefficients of many terms that is bounded for biholomorphic mappings from certain domains in $\mathbb{C}^n$. FitzGerald's [3] following counter-example tells us that there is no such combination. Actually, for any combination of the coefficients of any terms of the Taylor expansion of a biholomorphic mapping in any domain in $\mathbb{C}^n$, the modulus of the combination is unbounded.

In one variable theory, the only normalized (meaning $f(0) = 0$, and $f'(0) = 1$) univalent analytic function on the whole z-plane is z. But there are many normalized biholomorphic mappings taking the space $\mathbb{C}^n$ into itself (cf. Rosay and Rudin[1]).

Let $F = (f_1, f_2, \cdots, f_n) : \mathbb{C}^n \to \mathbb{C}^n$ be a normalized holomorphic mapping on $\mathbb{C}^n$, i.e., $F(0) = 0$ and $J_F(0) = I$ (I is the identity matrix). Then each component of F is a holomorphic function of several complex variables $z = (z_1, z_2, \cdots, z_n)$, and can be written as follows:

$$f_k(z_1, z_2, \cdots, z_n) = z_k + \sum d^{(k)}_{(j_1, j_2, \cdots, j_n)} z_1^{j_1} z_2^{j_2} \cdots z_n^{j_n}$$

where $k = 1, 2, \cdots, n$ and each j_m, $m = 1, 2, \cdots, n$ is a non-negative integer, and $j_1 + j_2 + \cdots + j_n \geq 2$.

The following two examples are biholomorphic mappings on $\mathbb{C}^n$.

Example 1. Let $b = (b_1, b_2, \cdots, b_n) \in \mathbb{C}^n$, $c = (c_1, c_2, \cdots, c_n) \in \mathbb{C}^n$, and define $b \cdot c = \sum_{i=1}^n b_i c_i$. Let $v \in \mathbb{C}^n$, and $v \neq 0$. Assume that $A, B, C, \cdots$ are vectors from $\mathbb{C}^n$ such that $A \cdot v = 0$, $B \cdot v = 0$, $C \cdot v = 0$, $\cdots$. Let a be a complex number. Consider the normalized polynomial mapping

$$w = z + av(A \cdot z)(B \cdot z)(C \cdot z) \cdots.$$

The product is finite and has at least two factors that involve the product of $(A \cdot z)$ and $(B \cdot z)$. Then this mapping is biholomorphic.

To prove the claim, it is sufficient to obtain the inverse of the mapping. Dot the equation with A. Since $A \cdot v = 0$, the equation is $A \cdot w = A \cdot z$.

Similarly, $B \cdot w = B \cdot z$, $C \cdot w = C \cdot z$, $\cdots$. Hence, $z = w - av(A \cdot w)(B \cdot w)(C \cdot w)\cdots$, and the mapping is inverted.

Example 2. Let a be a nonzero complex number and define a biholomorphic mapping of $\mathbb{C}^n$ into $\mathbb{C}^n$ by its coordinate functions: $w_1 = z_1 \exp(az_2)$, and $w_k = z_k$ for $k = 2, 3, \cdots, n$.

Clearly, $z_k = w_k$ for $k = 2, 3, \cdots, n$, and $\exp(az_2)$ is known and is nonzero. Thus $z_1 = w_1 \exp(-aw_2)$. We obtain the inverse of the mapping.

Applying a permutation of the independent variables and the same permutation of the dependent variables, it is possible to create other normalized biholomorphic mappings of $\mathbb{C}^n$ into $\mathbb{C}^n$ from the previous examples. For the discussion here, it is important to see how the lower order terms behave under the composition of these examples.

Let $m \geq 1$ be an integer. If $w = z + P_m(z) + O(|z|^{m+1})$ and $w = z + Q_m(z) + O(|z|^{m+1})$ are two mappings where P_m and Q_m are vectors in $\mathbb{C}^n$ with each coordinate function being a homogeneous polynomial of degree m and all other terms of high order indicated by the expression $O(|z|^{m+1})$, then the composition of these mappings is given by

$$w = z + P_m(z) + Q_m(z) + O(|z|^{m+1}).$$

Examples 1 and 2 show that there are many biholomorphic mappings of $\mathbb{C}^n$ into $\mathbb{C}^n$. There are many coefficients of second order terms. For each coordinate function there are $\dfrac{n(n+1)}{2}$ coefficients; for the full mapping, there are $\dfrac{n^2(n+1)}{2}$ coefficients of second order terms. We already know that the magnitude of each coefficient is unbounded. But it is still possible that the magnitude of some combination of coefficients is bounded. The striking fact is that there is no limitation on choice of the coefficients of second order terms! Given a set of complex numbers for the respective coefficients of second order expressions, there is a way to extend the multivariable power series such that the resulting map is defined and biholomorphic on $\mathbb{C}^n$.

Theorem 5.1.1 (Fitzgerald). *Let $\{P_1, P_2, \cdots, P_n\}$ be a sequence of n homogeneous polynomials of second order in n variables $(n \geq 2)$. Then, for each $k = 1, 2, \cdots, n$, there exists a function*

$$f_k(z_1, z_2, \cdots, z_n) = z_k + P_k(z_1, z_2, \cdots, n) + O(|z|^3)$$

such that $F = (f_1, f_2, \cdots, z_n)$ is a biholomorphic mapping of $\mathbb{C}^n$ into $\mathbb{C}^n$.

Proof. To generate all the second order terms, we need only to consider the following four biholomorphic mappings of $\mathbb{C}^n$ into $\mathbb{C}^n$, where we write the expansion only up to second order.

$$
\begin{aligned}
(1) w_1 &= z_1 + az_1^2, \\
w_2 &= z_2, \\
&\cdots, \\
w_n &= z_n.
\end{aligned}
\qquad\qquad
\begin{aligned}
(2) w_1 &= z_1 + az_1 z_2, \\
w_2 &= z_2, \\
&\cdots, \\
w_n &= z_n.
\end{aligned}
$$

$$
\begin{aligned}
(3) w_1 &= z_1 + az_2^2, \\
w_2 &= z_2, \\
&\cdots, \\
w_n &= z_n.
\end{aligned}
\qquad\qquad
\begin{aligned}
(4) w_1 &= z_1 + az_2 z_3, \\
w_2 &= z_2, \\
&\cdots, \\
w_n &= z_n.
\end{aligned}
$$

Consider a permutation on the set $\{2, 3, \cdots, n\}$. Apply the same permutation to the indices of both the independent and dependent variables. Each second order term for the first coordinate function can be obtained in this way. By permutating $\{1, 2, \cdots, n\}$, every second order term in any coordinate function can be obtained, from (1) through (4).

These four initial segments of mappings would generate all possible segments up to second order using permutations of both the independent and dependent variables and by compositions. It suffices to show that these four initial segments are indeed the initial segments of normalized biholomorphic mappings of $\mathbb{C}^n$ into $\mathbb{C}^n$. In cases (3) and (4), these are such biholomorphic mappings. In the case of (2), this is the initial segment of example 2.

It remains only to find an appropriate type of mapping which has (1) for its initial segment. In Example 1, we consider $v = (1, 1, 0, \cdots, 0)$ and $A = B = (1, -1, 0, \cdots, 0)$. The mapping is

$$
\begin{aligned}
w_1 &= z_1 + a(z_1 - z_2)^2 = z_1 + az_1^2 - 2az_1 z_2 + az_2^2, \\
w_2 &= z_2 + a(z_1 - z_2)^2 = z_2 + az_1^2 - 2az_1 z_2 + az_2^2, \\
w_3 &= z_3, \\
&\cdots, \\
\text{(5.1.1)} \qquad w_n &= z_n.
\end{aligned}
$$

Consider a mapping with initial segment (3) with a replaced by $-a$.

$$w_1 = z_1 - az_2^2,$$
$$w_2 = z_2,$$
$$\cdots,$$
(5.1.2)
$$w_n = z_n.$$

The composition of (5.1.1) and (5.1.2) is the following biholomorphic mapping:

$$w_1 = z_1 + az_1^2 - 2az_1z_2,$$
$$w_2 = z_2 + az_1^2 - 2az_1z_2 + az_2^2,$$
$$w_3 = z_3,$$
$$\cdots,$$
(5.1.3)
$$w_n = z_n.$$

Again consider a mapping with initial segment (3) with a replaced by $-a$. Now exchange indices 1 and 2 in the subscripts of the independent and dependent variables.

$$w_1 = z_1,$$
$$w_2 = z_2 - az_1^2,$$
$$w_3 = z_3,$$
$$\cdots,$$
(5.1.4)
$$w_n = z_n.$$

The composition of (5.1.3) and (5.1.4) is the following biholomorphic mapping:

$$w_1 = z_1 + az_1^2 - 2az_1z_2,$$
$$w_2 = z_2 - 2az_1z_2 + az_2^2,$$
$$w_3 = z_3,$$
$$\cdots,$$
(5.1.5)
$$w_n = z_n.$$

Consider mapping (2) with a replaced by $2a$,

$$w_1 = z_1 + 2az_1z_2,$$
$$w_2 = z_2,$$
$$\cdots,$$

(5.1.6) $$w_n = z_n.$$

The composition mapping of (5.1.5) and (5.1.6) is the following biholomorphic mapping:

$$w_1 = z_1 + az_1^2 + O(|z|^3),$$
$$w_2 = z_2 + az_2^2 - 2az_1z_2,$$
$$w_3 = z_3,$$
$$\cdots,$$

(5.1.7) $$w_n = z_n.$$

Consider a mapping with initial segment (2) with a replaced by $2a$ and with 1 and 2 interchanged in the subscripts of both the independent variable and dependent variable.

$$w_1 = z_1,$$
$$w_2 = z_2 + 2az_1z_2 + O(|z|^3),$$
$$w_3 = z_3,$$
$$\cdots,$$

(5.1.8) $$w_n = z_n.$$

The composition of (5.1.7) and (5.1.8) is the following biholomorphic mapping:

$$w_1^{\cdot} = z_1 + az_1^2 + O(|z|^3),$$
$$w_2 = z_2 + az_2^2 + O(|z|^3),$$
$$w_3 = z_3,$$
$$\cdots,$$

(5.1.9) $$w_n = z_n.$$

In a similar fashion, in Example 1, consider $v = (1, -1, 0, \cdots, 0)$ and $A = B = (1, 1, 0, \cdots, 0)$. The mapping is

$$w_1 = z_1 + a(z_1 + z_2)^2,$$
$$w_2 = z_2 - a(z_1 - z_2)^2,$$
$$w_3 = z_3,$$
$$\cdots,$$

(5.1.10) $$w_n = z_n.$$

After a similar reduction, the final composition mapping is

$$w_1 = z_1 + az_1^2 + O(|z|^3),$$
$$w_2 = z_2 - az_2^2 + O(|z|^3),$$
$$w_3 = z_3,$$
$$\cdots,$$

(5.1.11) $\qquad\qquad\qquad w_n = z_n.$

The composition of the mappings (5.1.9) and (5.1.11) results in the mapping

$$w_1 = z_1 + 2az_1^2 + O(|z|^3),$$
$$w_2 = z_2 + O(|z|^3),$$
$$w_3 = z_3,$$
$$\cdots,$$

(5.1.12) $\qquad\qquad\qquad w_n = z_n.$

After a is replaced by $\dfrac{1}{2}a$, (5.1.12) has the desired initial segment (1).

Hence the theorem has been proved.

In classical geometric function theory, "coefficient body" means the set of points in $\mathbb{C}^m$ such that the coordinates of the point are the first few coefficients of the power series of a function in S starting with the coefficients of the second order term: $(a_2, a_3, \cdots, a_{m+1})$. A major reference for this subject is Schaeffer and Spencer[1]; particularly Plates I and II which show the coefficient body for the coefficients of the second and third order terms. Given this background, it is reasonable to extend Theorem 5.1.1. to include the coefficients of all second and third order terms. It shown that the corresponding "coefficient body" for higher dimensions is the whole space.

We consider only the normalized mapping $w_1, w_2, \cdots, w_n$, and we need only to consider the mappings for which all the second order coefficients are zero. To generate all the third order terms, we need only to consider the following seven holomorphic mappings of $\mathbb{C}^n$ into $\mathbb{C}^n$, where we write the expansion only up to third order:

$$(1) \quad \begin{aligned} w_1 &= z_1 + az_1^3, \\ w_2 &= z_2, \\ &\cdots, \\ w_n &= z_n. \end{aligned} \qquad (2) \quad \begin{aligned} w_1 &= z_1 + az_1^2 z_2, \\ w_2 &= z_2, \\ &\cdots, \\ w_n &= z_n. \end{aligned} \qquad (3) \quad \begin{aligned} w_1 &= z_1 + az_1 z_2^2, \\ w_2 &= z_2, \\ &\cdots, \\ w_n &= z_n. \end{aligned}$$

$$(4) \quad \begin{aligned} w_1 &= z_1 + az_1 z_2 z_3, \\ w_2 &= z_2, \\ &\cdots, \\ w_n &= z_n. \end{aligned} \qquad (5) \quad \begin{aligned} w_1 &= z_1 + az_2^3, \\ w_2 &= z_2, \\ &\cdots, \\ w_n &= z_n. \end{aligned} \qquad (6) \quad \begin{aligned} w_1 &= z_1 + az_2^2 z_3, \\ w_2 &= z_2, \\ &\cdots, \\ w_n &= z_n. \end{aligned}$$

$$(7) \quad \begin{aligned} w_1 &= z_1 + az_2 z_3 z_4 \\ w_2 &= z_2 \\ &\cdots, \\ w_n &= z_n. \end{aligned}$$

Consider permutations of the set $\{2, 3, \cdots, n\}$. Apply the same permutation to the indices of both the independent and dependent variables. Each third order term for the first coordinate function can be obtained in this way. By permuting $\{1, 2, \cdots, n\}$, every third order term in any coordinate function can be obtained from (1) through (7).

These seven initial segments of mappings would generate all possible segments up to third order using permutations of both the independent and dependent variables and by compositions. It suffices to show that these seven initial segments are indeed the initial segments of normalized biholomorphic mappings of $\mathbb{C}^n$ into $\mathbb{C}^n$.

Initial segments (5), (6) and (7) are normalized biholomorphic mappings of the type of example 1: Specifically, the choice $v = (1, 0, 0, \cdots, 0)$ and $A = B = C = (0, 1, 0, \cdots, 0)$ gives the initial segment (5). The selection $v = (1, 0, 0, \cdots, 0)$ and $A = B = (0, 1, 0, \cdots, 0)$, $C = (0, 0, 1, 0, \cdots, 0)$ gives the initial segment (6), and the specification that $v = (1, 0, 0, \cdots, 0)$ and $A = (0, 1, 0, \cdots, 0)$, $B = (0, 0, 1, 0, \cdots, 0)$, $C = (0, 0, 0, 1, 0, \cdots, 0)$ gives initial segment (7).

Initial segment (4) arises as an extension of example 2: $w_1 = z_1 \exp(az_2 z_3)$, $w_2 = z_2$, $\cdots$, $w_n = z_n$. This is a biholomorphic mapping. Given $w_1, w_2, \cdots, w_n$, the values of $z_2, z_3, \cdots, z_n$ can be obtained immediately. The value of $z_2 z_3$ is then determined. Since the exponential is never equal to zero, the value of z_1 can be computed.

Initial segment (3) arises as an extension of Example 2: $w_1 = z_1 \exp(az_2^2)$, $w_2 = z_2$, $\cdots$, $w_n = z_n$.

To obtain an initial segment of form (1), the compositions of several mappings will be considered. In example 1, let $v = (1, -1, 0, \cdot, 0)$ and $A = B = C = (1, 1, 0, \cdots, 0)$.

$$w_1 = z_1 + a(z_1 + z_2)^3,$$
$$w_2 = z_2 - a(z_1 + z_2)^3,$$
$$w_3 = z_3,$$
$$\cdots,$$

(5.1.13) $$w_n = z_n.$$

If in (5), we replace a by $-a$, and in (3), we replace a by $-3a$, then compose them with (5.1.13), we have

$$w_1 = z_1 + az_1^3 + 3az_1^2 z_2 + O(|z|^4),$$
$$w_2 = z_2 - a(z_1 + z_2)^3,$$
$$w_3 = z_3,$$
$$\cdots,$$

(5.1.14) $$w_n = z_n.$$

Exchanging z_1 and z_2 in (3) and (5), replacing a by $3a$ in (3), composing with (5.1.14), and then composing with (5), we have

$$w_1 = z_1 + az_1^3 + 3az_1^2 z_2 + O(|z|^4),$$
$$w_2 = z_2 - 3az_1 z_2^2 - az_2^3 + O(|z|^4),$$
$$w_3 = z_3,$$
$$\cdots,$$

(5.1.15) $$w_n = z_n.$$

In example 1, let $v = (1, 1, 0, \cdots, 0)$ and let $A = B = C = (1, -1, 0, \cdots, 0)$, and we obtain

$$w_1 = z_1 + a(z_1 - z_2)^3,$$
$$w_2 = z_2 + a(z_1 - z_2)^3,$$
$$w_3 = z_3,$$
$$\cdots,$$

(5.1.16) $$w_n = z_n.$$

Apply the same process as we used from (5.1.13) to (5.1.15), and we get a biholomorphic mapping

$$w_1 = z_1 + az_1^3 - 3az_1^2 z_2 + O(|z|^4),$$
$$w_2 = z_2 + 3az_1 z_2^2 - az_2^3 + O(|z|^4),$$
$$w_3 = z_3,$$
$$\cdots,$$

(5.1.17) $$w_n = z_n.$$

Compose (5.1.15) with (5.1.17), and we have a biholomorphic mapping

$$w_1 = z_1 + 2az_1^3 + O(|z|^4),$$
$$w_2 = z_2 - 2az_2^3 + O(|z|^4),$$
$$w_3 = z_3,$$
$$\cdots,$$

(5.1.18) $$w_n = z_n.$$

Similar calculations can be done with $v = (2, -1, 0, \cdots, 0)$ and $A = B = C = (1, 2, 0, \cdots, 0)$ for an analogue of this process from (5.1.14) to (5.1.15). Then use $v = (2, 1, 0, \cdots, 0)$ and $A = B = C = (1, -2, 0, \cdots, 0)$ for the analogue of this process from (5.1.16) to (5.1.17). Finally, the composition gives the following analogue of (5.1.18):

$$w_1 = z_1 + 4az_1^3 + O(|z|^4),$$
$$w_2 = z_2 - 16az_2^3 + O(|z|^4),$$
$$w_3 = z_3,$$
$$\cdots,$$

(5.1.19) $$w_n = z_n.$$

In (5.1.18), we replace a by $-8a$, then composing with (5.1.19), we have a biholomorphic mapping

$$w_1 = z_1 - 12az_1^3 + O(|z|^4),$$
$$w_2 = z_2 + O(|z|^4),$$
$$w_3 = z_3,$$
$$\cdots,$$

(5.1.20) $$w_n = z_n.$$

We obtain a biholomorphic mapping in the form (1) if we replace a by $-\dfrac{1}{12}a$.

The final initial segment (2) is easily created. Consider the (not normalized) biholomorphic mapping

$$w_1 = z_1 + z_2,$$
$$w_2 = z_2,$$
$$w_3 = z_3,$$
$$\cdots,$$

(5.1.21)
$$w_n = z_n.$$

Composing (5.1.21) with (1), we obtain a biholomorphic mapping

$$w_1 = z_1 + z_2 + a(z_1 + z_2)^3 + O(|z|^4),$$
$$w_2 = z_2 + O(|z|^4),$$
$$w_3 = z_3,$$
$$\cdots,$$

(5.1.22)
$$w_n = z_n.$$

Apply (5.1.22) in the same process as we used for (5.1.13) to (5.1.14), and we have a biholomorphic mapping

$$w_1 = z_1 + z_2 + az_1^3 + 3az_1^2 z_2 + O(|z|^4),$$
$$w_2 = z_2 + O(|z|^4),$$
$$w_3 = z_3,$$
$$\cdots,$$

(5.1.23)
$$w_n = z_n.$$

Consider a biholomorphic mapping in the form (1) as

$$w_1 = z_1 - az_1^3 + O(|z|^4),$$
$$w_2 = z_2 + O(|z|^4),$$
$$w_3 = z_3,$$
$$\cdots,$$

(5.1.24)
$$w_n = z_n.$$

Composing (5.1.23) and (5.1.24), we obtain a biholomorphic mapping

$$w_1 = z_1 + z_2 + 3az_1^2 z_2 + O(|z|^4),$$
$$w_2 = z_2 + O(|z|^4),$$
$$w_3 = z_3,$$
$$\cdots,$$

(5.1.25) $$w_n = z_n.$$

Composing (5.1.25) and the inverse mapping of (5.1.21), we have

$$w_1 = z_1 + 3az_1^2 z_2 + O(|z|^4),$$
$$w_2 = z_2 + O(|z|^4),$$
$$w_3 = z_3,$$
$$\cdots,$$

(5.1.26) $$w_n = z_n.$$

Replace a by $\dfrac{a}{3}$, and we obtain a biholomorphic mapping with initial segment (2). Thus the full set of generators has been found. If the coefficients of the second order terms are required to all be zero, then the coefficients of the third terms can be chosen arbitrarily. Finally, it is shown that all the second and third order coefficients can be chosen arbitrarily.

Theorem 5.1.2. *Let* $\{P_1, P_2, \cdots, P_n\}$ *be a sequence of* n *polynomials having only second and third order terms in* n *variables. Then, for each* $k = 1, 2, \cdots, n$, *there exists a function*

$$f_k(z_1, z_2, \cdots, z_n) = z_k + P_k(z_1, z_2, \cdots, z_n) + O_k(|z|^4)$$

such that $F = (f_1, f_2, \cdots, f_n)$ *is a biholomorphic mapping of* $\mathbb{C}^n$ *into* $\mathbb{C}^n$.

Proof. Let $\mathbb{F}$ be a mapping of $\mathbb{C}^n$ into $\mathbb{C}^n$ with its k-th coordinate function being $z_k + P_k(z_1, z_2, \cdots, z_n)$. Then $\mathbb{F}$ may not be biholomorphic. By Theorem 5.1.1, there exists a normalized biholomorphic mapping G of $\mathbb{C}^n$ into $\mathbb{C}^n$ such that G has the same second order coefficients as $\mathbb{F}$. Then G^{-1} has exactly the opposite sign for each of the coefficients of the second order terms. In the mapping $G^{-1}(\mathbb{F}(z))$, the coefficient of each second order term is zero. By the discussion after Theorem 5.1.1, there exists a normalized mapping H of $\mathbb{C}^n$ into $\mathbb{C}^n$ such that the coefficient of each second order term is zero and the coefficient of each third order term is exactly the same as that of $G^{-1}(\mathbb{F}(z))$. Then H^{-1} has the opposite sign for each third order coefficient. Then the mapping $H^{-1}(G^{-1}(\mathbb{F}(z)))$ has no nonzero second or third order terms. Consider the composition $G(H(z))$. It is a normalized biholomorphic mapping of $\mathbb{C}^n$ into $\mathbb{C}$ and agrees with $\mathbb{F}$ up to third order. Hence, the composition is a mapping of the form the theorem claims to exist.

Theorem 5.1.2. can be extended. It should be noted that Andersén[1] and Andersén and Lempert[1] have discussed related questions for volume-preserving mappings.

As a consequence of theorems 5.1.1. and 5.1.2, the normalized biholomorphic mappings of $\mathbb{C}^n$ do not form a normal family. This situation strongly suggests requiring some additional property of the mappings of the family to form a normal family.

Theorems 5.1.1 and 5.1.2 also strongly suggest requiring some additional property of mappings of a family in order to get some positive results. For example, the convexity and the starlikeness are very natural additional properties.

§5.2 Convex mappings

Let $\Omega \subset \mathbb{C}^n$ be a domain in $\mathbb{C}^n$ with $0 \in \Omega$. All normalized biholomorphic mappings on Ω form a family. We denote it by $S(\Omega)$. We already know that there is no Bieberbach conjecture for the family $S(\Omega)$. However, we still can get some estimate of the modulus of coefficients of the Taylor expansion of the mappings of some subfamilies of $S(\Omega)$.

Let $\Omega \subset \mathbb{C}^n$ be a domain. The holomorphic mapping $f = (f_1, \cdots, f_n)$ maps Ω into $\mathbb{C}^n$, f is called a *convex mapping* if for any $a \in \Omega, b \subset \Omega$, the chord connects $f(a)$ and $f(b)$ lies in $f(\Omega)$.

In this section, we study the family of convex mappings.

Let us consider the Reinhardt domain

$$(5.2.1) \quad B_p = \left\{ z = (z_1, \cdots, z_n) \in \mathbb{C}^n : \|z\|_p = \left(\sum_{i=1}^{n} |z_i|^p \right)^{\frac{1}{p}} < 1 \right\}, \quad p \geq 1$$

then we have the following results (cf. Gong [6], Gong-Liu [1], FitzGerald [4]):

Theorem 5.2.1. *Let B_p be defined by (5.2.1), and let $f(z) : B_p \to \mathbb{C}^n$ be a normalized biholomorphic convex mapping on B_p. The expansion of*

$f(z) = (f_1(z), \cdots, f_n(z))$ *can be expressed as*

$$(5.2.2) \qquad\qquad f(z) = z + \sum_{k=2}^{\infty} \varphi^{(k)}(z),$$

where $\varphi^{(k)}(z) = (\varphi_1^{(k)}(z), \cdots, \varphi_n^{(k)}(z))$ *and* $\varphi_j^{(k)}(z)$ *for* $j = 1, 2, \cdots, n$ *are homogeneous polynomials of the elements of* $z = (z_1, \cdots, z_n)$ *of degree* k. *Then*

$$(5.2.3) \qquad\qquad \|\varphi^{(k)}(z)\|_p \le \|z\|_p^k$$

holds for all $k = 2, 3, 4, \cdots$, *where*

$$\|\varphi^{(k)}(z)\|_p = \left(\sum_{j=1}^{n} |\varphi_j^{(k)}(z)|^p \right)^{\frac{1}{p}}.$$

Corollary 5.2.1. *Let* $B^n = \{z = (z_1, \cdots, z_n) \in \mathbb{C}^n : \sum_{i=1}^{n} |z_i|^2 < 1\}$ *be the unit ball in* $\mathbb{C}^n$, *and let* $f(z): B^n \to \mathbb{C}^n$ *be a normalized biholomorphic convex mapping on* B^n. *The expansion of* $f(z)$ *is* (5.2.2), *then*

$$(5.2.4) \qquad\qquad \sum_{j=1}^{n} |\varphi_j^{(k)}(z)|^2 \le \left(\sum_{j=1}^{n} |z_j|^2 \right)^k$$

holds for $k = 2, 3, 4, \cdots$.

Let $\mathbb{P}^n = \{z = (z_1, \cdots, z_n) \in \mathbb{C}^n : |z_j| < 1, j = 1, \cdots, n\}$ be the unit polydisk in $\mathbb{C}^n$, and $f(z) = (f_1(z), \cdots, f_n(z)): P^n \to \mathbb{C}^n$ be a normalized biholomorphic convex mapping on $\mathbb{P}^n$. Suffridge (T. J. Suffridge [1]) proved the following theorem:

Theorem 5.2.2. *Let* $f : \mathbb{P}^n \to \mathbb{C}^n$ *be a locally biholomorphic mapping on* $\mathbb{P}^n$, *then* f *is biholomorphic convex mapping if and only if there exists* n *univalent convex functions* $f_1, f_2, \cdots, f_n$ *on the unit disc* D, *such that*

$$f(z) = (f_1(z_1), \ f_2(z_2), \ \cdots, \ f_n(z_n))T$$

holds where T *is a nonsingular* $n \times n$ *constant matrix.*

Thus, if $f(z)$ is a normalized biholomorphic convex mapping on $\mathbb{P}^n$, then the Taylor expansion at $z = 0$ is

$$f(z) = z + \sum_{k=1}^{\infty} (\varphi_1^{(k)}(z), \cdots, \varphi_n^{(k)}(z))$$

where $\varphi_j^{(k)}(z) = \sum\limits_{i=1}^{n} a_{ij} z_i^k$, $\quad j = 1, \cdots, n$, $\quad k = 2, 3, \cdots$. Moreover, $\|z\|_\infty = \max\limits_{j} |z_j|$. Thus, we have

Corollary 5.2.2. *Let $\mathbb{P}^n$ be the unit polydisk in $\mathbb{C}^n$, and let $f(z) : \mathbb{P}^n \to \mathbb{C}^n$ be a normalized biholomorphic convex mapping on $\mathbb{P}^n$. The expansion of $f(z)$ is (5.2.2), and $\varphi_j^k(z) = \sum\limits_{i=1}^{n} a_{ij} z_i^k$, $j = 1, \cdots, n, k = 2, 3, \cdots$, then*

$$\max_{j} \left| \sum_{i=1}^{n} a_{ij} z_i^k \right| \leq \max_{j} |z_j|^k$$

holds for all $k = 2, 3, 4, \cdots$.

Of course, Corollary 5.2.1 and 5.2.2 are two special cases of Theorem 5.2.1 where $p = 2$ and $p = \infty$. When $n = 1$, Corollary 5.2.1 and 5.2.2 coincide with the classical result (Theorem 1.1.5 (1.1.23)): If $f(z) = z + \sum\limits_{n=2}^{\infty} a_n z^n$ is a univalent convex holomorphic function, then $|a_n| \leq 1$ for $n = 2, 3, \cdots$.

Proof of Theorem 5.2.1.

Let k be a positive integer, $g_k(z) = \frac{1}{k} \sum\limits_{j=1}^{k} f(e^{\frac{i z j \pi}{k}} z)$, then $g_k(z)$ is majorant with $f(z)$ since $f(z)$ is convex, i.e., $g_k(z) \prec f(z)$. Let $\nu^{(k)}(z) = f^{-1}(g_k(z)) = (\nu_1^{(k)}(z), \cdots, \nu_n^{(k)}(z))$, then $\nu^{(k)}(z)$ is a holomorphic mapping on B_p, which maps B_p into B_p with $\nu^{(k)}(0) = 0$ and

(5.2.5) $g_k(z) = f(\nu^{(k)}(z)).$

By the definition of $g_k(z)$ and $\nu^{(k)}(z)$, we have

(5.2.6) $g_k(z) = \sum\limits_{j=1}^{\infty} \varphi^{(kj)}(z)$

and

(5.2.7) $f(\nu^{(k)}(z)) = \nu^{(k)}(z) + \sum\limits_{j=2}^{\infty} \varphi^{(j)}(\nu^{(k)}(z)).$

Substituting (5.2.6) and (5.2.7) into (5.2.5), and comparing the lowest degree term, we have

$$\varphi^{(k)}(z) = \text{ the lowest degree term of } \nu^{(k)}(z) = \frac{1}{2\pi} \int_0^{2\pi} \nu^{(k)}(e^{i\theta} z) e^{-ik\theta} d\theta,$$

and then

$$|\varphi_j^{(k)}(z)| \leq \left(\frac{1}{2\pi}\int_0^{2\pi}|\nu_j^{(k)}(e^{i\theta})|^p d\theta\right)^{\frac{1}{p}}$$

since $p \geq 1$. Then

$$\sum_{j=1}^n|\varphi_j^{(k)}(z)|^p \leq \frac{1}{2\pi}\int_0^{2\pi}\sum_{j=1}^n|\nu_j^{(k)}(e^{i\theta}z)|^p d\theta < 1.$$

Let $\|\varphi^{(k)}(z)\|_p = \left(\sum_{j=1}^n|\varphi_j^{(k)}(z)|^p\right)^{\frac{1}{p}}$, then the above inequality implies that $\|\varphi^{(k)}(z)\|_p < 1$ holds for all $z \in B_p$.

If $z \in B_p$, then $\frac{z}{\|z\|_p} \in \bar{B}_p$, and

$$\|\varphi^{(k)}(z)\|_p^p = \sum_{j=1}^n|\varphi_j^{(k)}(z)|^p = \sum_{j=1}^n\|z\|_p^{kp}\left|\varphi_j^{(k)}\left(\frac{z}{\|z\|_p}\right)\right|^p \leq \|z\|_p^{kp}$$

We have (5.2.3) for all $k = 2, 3, \cdots$.

Moreover, since

$$\|f(z)\|_p \leq \|z\|_p + \sum_{k=2}^\infty\|\varphi^{(k)}(z)\|_p,$$

we have

(5.2.8)
$$\|f(z)\|_p \leq \|z\|_p + \sum_{k=2}^\infty\|z\|_p^k = \frac{\|z\|_p}{1 - \|z\|_p}$$

by (5.2.3).

Actually, we can prove the following result. (T.J.Suffridge[2], C.R.Thomas[1], T.S.Liu[4]).

Theorem 5.2.3. *If $f(z)$ is a normalized biholomorphic convex mapping on the unit ball B^n in C^n, then*

(5.2.9)
$$\frac{|z|}{1+|z|} \leq |f(z)| \leq \frac{|z|}{1-|z|}.$$

This estimate is precise. There are many mappings for which the equalities hold.

Proof. The right hand side inequality of (5.2.9) is a special case of (5.2.8) when $p = 2$. We only need to prove the left side inequality of (5.2.9).

Let $z = (z_1, \cdots, z_n) \in B^n$, and $z(t) = f^{-1}(tf(z))$, $0 \le t \le 1$. Then $z(t)$ is well-defined since f is convex. We have $f(z(t)) = tf(z)$ and

$$(5.2.10) \qquad \frac{df(z(t))}{dt} = f(z) = \frac{1}{t}f(z(t)).$$

Moreover,

$$\frac{df(z(t))}{dt} = f'(z(t))\frac{dz(t)}{dt}.$$

Hence

$$(5.2.11) \qquad \frac{1}{t}f(z(t)) = f'(z(t))\frac{d(z(t))}{dt}.$$

Let $\varphi_a(z) \in Aut(B^n)$, $\varphi_a(0) = a$, where $Aut(B^n)$ is the group of holomorphic automorphisms of B^n(cf. Hua[1]). Then $f(\varphi_a(z))$ is a convex mapping. Expand $f(\varphi_a(z))$ as a Taylor series of z at a neighborhood of $z = 0$,

$$f(\varphi_a(z)) = f(a) + J_f(a)J_{\varphi_a}(0)z + \cdots$$

where f and z are column vectors. Let

$$(5.2.12) \qquad h_a(z) = (J_{\varphi_a}(0))^{-1}(J_f(a))^{-1}(f(\varphi_a(z)) - f(a)),$$

then $h_a(z) = z + \cdots$. This implies $h_a(z)$ is a normalized convex biholomorphic mapping. By the right inequality of (5.2.9), we have

$$(5.2.13) \qquad |h_a(z)| \le \frac{|z|}{1 - |z|}.$$

Replacing a and z by $z(t)$ in (5.2.12), we have

$$(5.2.14) \qquad h_{z(t)}(z(t)) = (J_{\varphi_{z(t)}}(0))^{-1}(J_f(z(t))^{-1}(-f(z(t)))).$$

From the expression of $\varphi_{z(t)}$, we obtain

$$(J_{\varphi_{z(t)}}(0))^{-1} = J_{\varphi_{z(t)}}(z(t)) = -\frac{A}{1 - |z(t)|^2}$$

where $A = sI + \dfrac{1}{1+s} z(t)\overline{z(t)}'$, $s^2 = 1 - |z(t)|^2$. Using (5.2.11), (5.2.14) can be rewritten as

$$h_{z(t)}(z(t)) = \frac{At}{1 - |z(t)|^2} \frac{dz(t)}{dt},$$

and hence

$$\overline{z(t)}' h_{z(t)}(z(t)) = \frac{\overline{z(t)}' A}{1 - |z(t)|^2} \frac{dz(t)}{dt} = \frac{\overline{z(t)}' t}{1 - |z(t)|^2} \frac{dz(t)}{dt}.$$

Taking the real part on both sides of the preceding equality, we obtain

$$2\Re(z(t)' h_{z(t)}(z(t))) = \frac{t}{1 - |z(t)|^2} 2\Re\left\{ z(t)' \frac{dz(t)}{dt} \right\}$$

$$= \frac{2t}{1 - |z(t)|^2} |z(t)| \frac{d|z(t)|}{dt}.$$

Using (5.2.13), we have

(5.2.15) $$\qquad \frac{1}{t} \geq \frac{1}{|z(t)|(1 + |z(t)|)} \frac{d|z(t)|}{dt}.$$

We know that

$$\frac{d|f(z(t))|^2}{dt} = 2|f(z(t))| \frac{d|f(z(t))|}{dt}$$

and

$$\frac{d|f(z(t))|^2}{dt} = 2\Re\left(\overline{f(z(t))}' \frac{df(z(t))}{dt} \right) = \frac{2}{t} |f(z(t))|^2$$

by (5.2.10). Combining these two equalities, we have

$$\frac{1}{|f(z(t)|} \frac{d|f(z(t)|}{dt} \geq \frac{1}{|z(t)|(1 + |z(t)|)} \frac{d|z(t)|}{dt}$$

by (5.2.15). Integrating both sides of the preceding inequality with respect to t from φ to 1, we obtain

$$\log |f(z)| - \log |f(z(\varphi))| \geq \log \frac{|z|}{1 + |z|} - \log \frac{|z(\varphi)|}{1 + |z(\varphi)|}$$

since $z(1) = z$. Letting $\varphi \to 0$, we obtain $\log |f(z)| \geq \log \dfrac{|z|}{1 + |z|}$ since $\displaystyle\lim_{\varphi \to 0} \frac{|f(z(\varphi))|}{|z(\varphi)|} = 1$. This completes the proof of Theorem 5.2.3.

As a consequence of Theorem 5.2.3, we have the following covering theorem:

Corollary 5.2.3. *Let $f(z) : B^n \to \mathbb{C}^n$ be a normalized biholomorphic convex mapping on B^n. Then $f(B^n)$ contains $\frac{1}{2}B^n$, i.e. $f(B^n) \supseteq \frac{1}{2}B^n$.*

Obviously, (5.2.9) is the generalization of the classical theorem of convex mappings on the unit disk, Theorem 1.1.5 (1.1.25).

Moreover, we can prove the following more general results (S. Gong and T. S. Liu [1], T. S. Liu and G. B. Ren[1])

Theorem 5.2.4. *Let B_p be defined by (5.2.1), and let $f(z) : B_p \to \mathbb{C}^n$ be a normalized biholomorphic convex mapping on B_p, then*

$$(5.2.16) \qquad \frac{\|z\|_p}{1 + \|z\|_p} \leq \|f(z)\|_p \leq \frac{\|z\|_p}{1 - \|z\|_p}.$$

and

$$(5.2.17) \qquad \frac{|z|}{1 + \|z\|_p} \leq |f(z)| \leq \frac{|z|}{1 - \|z\|_p}$$

where $\|f(z)\|_p = \left(\sum_{i=1}^{n} |f_i(z)|(p) \right)^{\frac{1}{p}}$.

We already proved the right hand side inequality of (5.2.16). We omit the proof of the left hand side of (5.2.16) and (5.2.17).

For more discussion about the growth theorem for convex mappings refer to my monograph (S. Gong [6]).

Actually, excluding the growth theorem, there are already a series of results about the convex mappings in several complex variables.

For example, Theorem 5.2.2 mentioned the criterion for convexity for holomorphic mappings on the polydisk, there are a series interesting results about the criterion for convexity for holomorphic mappings in several complex variables (cf. S. Gong [6]).

The following theorem is the criterion for convexity for holomorphic mappings on the unit ball B^n in $\mathbb{C}^n$ (cf. S. Gong, S. K. Wang and Q. H. Yu [1] and K. Kikuchi [1]):

Theorem 5.2.5. *Let $f : B^n \to \mathbb{C}^n$ be a normalized locally biholomorphic mapping on B^n, then f is a normalized biholomorphic convex mapping on B^n if and only if the following condition is satisfied.*

Let $z \in B^n$ and $b = (b_1, b_2, \cdots, b_n) \in \mathbb{C}^n$ be any vector satisfying the condition $\Re\{z\bar{b}'\} = 0$, then

$$(5.2.18) \qquad |b|^2 + \Re\left\{ \sum_{i,j,\alpha,\beta,\gamma} b_\beta b_\gamma \frac{\partial w_i}{\partial z_\beta} \frac{\partial w_j}{\partial z_\gamma} \bar{z}_\alpha \frac{\partial^2 z_\alpha}{\partial w_i \partial w_j} \right\} \geq 0$$

holds, where $f(z) = (w_1, \cdots, w_n)$.

Of course, there are many other results about the criteria for convexity for holomorphic mappings. (cf. S. Gong [6]).

We omit the details of proof of Theorem 5.2.5.

Finally, we would like to state a result about the distortion theorem:

Let $\mathcal{S}_0$ be a family of normalized locally biholomorphic mappings on unit ball $B^n \subset \mathbb{C}^n$. $\mathcal{S}_0$ is called as a *linear-invariant family*, if for any $f \in \mathcal{S}_0$, and any $\varphi \in \mathrm{Aut}(B^n)$, where $\mathrm{Aut}(B^n)$ means the holomorphic automorphism group of B^n, the holomorphic mapping $g(z) = (f(\varphi(z)) - f(\varphi(0)))(J_{f \circ \varphi}(0))^{-1} \in \mathcal{S}_0$ again.

We may extend Theorem 1.1.1 as follows (cf. R. W. Barnard, C. H. FitzGerald and S. Gong [1], T. S. Liu [2]).

Theorem 5.2.6. *Let $\mathcal{S}_0$ be a linear-invariant family of normalized locally biholomorphic mappings on B^n. If $f \in \mathcal{S}_0$ and $f(z) = z + (zA^{(1)}z', \cdots, zA^{(n)}z') + \cdots$ where $A^{(i)} = (a^i_{jk})_{1 \leq j,k \leq n}, i = 1, 2, \cdots, n$ are constant matrices. Then for every $z \in B^n$, the inequality*

$$(5.2.19) \qquad \left| \log \frac{\det J_f(z)}{K(z, \bar{z})/K(0, 0)} \right| \leq C(\mathcal{S}_0) \log \frac{1 + |z|}{1 - |z|}$$

holds, where $K(z, \bar{z}) = \frac{n!}{\pi^n} \frac{1}{(1 - |z|^2)^{n+1}}$ is the Bergman kernel function of B^n, $C(\mathcal{S}_0) = \sup\{| \sum_{i=1}^{n} a^i_{ij}| : j = 1, \cdots, n, f \in \mathcal{S}_0\}$.

From Theorem 5.2.6 we immediately get the following consequence:

Corollary 5.2.4. *The assumptions are same as Theorem 5.2.6. Then for every $z \in B^n$, the inequality*

$$(5.2.20) \qquad \frac{(1 - |z|)^{c(\mathcal{S}_0) - \frac{n+1}{2}}}{(1 + |z|)^{c(\mathcal{S}_0) + \frac{n+1}{2}}} \leq |\det J_f(z)| \leq \frac{(1 + |z|)^{c(\mathcal{S}_0) - \frac{n+1}{2}}}{(1 - |z|)^{c(\mathcal{S}_0) + \frac{n+1}{2}}}$$

holds.

Theorem 5.2.6 and Corollary 5.2.4 can be extended to bounded symmetric domains. (cf. L. K. Hua [1], S. Gong and X. A. Zheng [1])

Of course, all normalized biholomorphic convex mappings on the unit ball B^n form a linear-invariant family. Thus (5.2.19) and (5.2.20) hold true for the family of normalized biholomorphic convex mappings on B^n. In this case, we may prove

$$\frac{n+1}{2} \leq C(\mathcal{S}_0) \leq 1 + \frac{\sqrt{2}(n-1)}{2}$$

(cf. R. W. Barnard, C. H. FitzGareld and S. Gong [1], T. S. Liu [2]). It was conjectured by Barnard, FitzGerald and Gong that $C(\mathcal{S}_0) = \frac{n+1}{2}$. Recently, Pfatzgraff and Suffridge (J. A. Pfatzgrall and T. J. Suffridge [1]) gave a counter-example to show that this conjecture is not true.

Question: What is the exact value of $C(\mathcal{S}_0)$ when $\mathcal{S}_0$ is the family of nomalized biholomorphic convex mappings on the unit ball B^n?

In my monograph (S. Gong [6]), there is more discussion about distortion theorems.

§5.3. Starlike mappings

In the previous section, we considered convex mappings including the estimate of the coefficients of convex mappings. In this section, we will discuss starlike mappings including the estimate of the coefficients of starlike mappings.

Let $\Omega \subset \mathbb{C}^n$ be a domain. The holomorphic mapping $f = (f_1, \cdots, f_n)$ maps Ω into $\mathbb{C}^n$ with $0 \in f(\Omega)$. f is called a *starlike mapping with respect to* 0 if for any $a \in \Omega$, the chord connecting 0 and $f(a)$ lies in $f(\Omega)$.

There are nearly no general results on the estimate of the modulus of the coefficients of the Taylor expansion of the family of the normalized starlike biholomorphic mappings. There are only a few results on unit polydisk.

Theorem 5.3.1. *Let $f(z) = z + \sum_{k=2}^{\infty} \varphi^{(k)}(z)$ be a normalized starlike bi-holomorphic mapping on the unit polydisk $\mathbb{P}^n$, where $f(z) = \begin{pmatrix} f_1(z) \\ \vdots \\ f_n(z) \end{pmatrix}, z = \begin{pmatrix} z_1 \\ \vdots \\ z_n \end{pmatrix}$ and $\varphi^{(k)}(z) = \begin{pmatrix} \varphi_1^{(k)}(z) \\ \vdots \\ \varphi_n^{(k)}(z) \end{pmatrix}, k = 2, 3, \cdots$ are column vectors, and*

$\varphi_j^{(k)}(z), j = 1, 2, \cdots, n$ are homogeneous polynomials of the elements of z of degree k, then

(5.3.1) $$\|\varphi^{(k)}(z)\| \le k\|z\|^k$$

holds for $k = 2, 3$, and $z \in \mathbb{P}^n$, where $\| \| = \| \|_\infty$ is the norm of the unit polydisk $\mathbb{P}^n$, that is $\|z\| = \|z\|_\infty = \max_j |z_j|$. (5.3.1) can be written as

(5.3.2) $$\max_j |\varphi_j^{(k)}(z)| \le k \max_j |z_j|^k.$$

When $k = 2$, this theorem was proved by M. Jahangiri [1], Y. P. Huang [1] and T. S. Liu [1]. When $k = 3$, this theorem was proved by T. S. Liu [1].

In order to prove this theorem, we need to use the following theorem of Suffridge [1].

Theorem 5.3.2.   Let $f(z) = z + \sum_{k=2}^\infty \varphi^{(k)}(z)$ be a normalized biholomorphic mapping on the unit polydisk $\mathbb{P}^n$, then $f(z)$ is starlike if and only if

(5.3.3) $$\Re\left\{\frac{w_i(z)}{z_i}\right\} \ge 0$$

holds for all $z \in \mathbb{P}^n$ and $i = 1, 2, \cdots, n$, where

(5.3.4) $$w(z) = \begin{pmatrix} w_1(z) \\ \vdots \\ w_n(z) \end{pmatrix} = J_f^{-1}(z)f(z) = z - \varphi^{(2)}(z) + \cdots,$$

$J_f(z)$ is the Jacobian of f at point z; $z, f(z)$ and $\varphi^{(k)}(z), k = 2, 3, \cdots$ are column vectors.

We omit the proof of Theorem 5.3.2.

Lemma 5.3.1.   Let $f(z) = z + \sum_{k=2}^\infty \varphi^{(k)}(z)$ be a normalized biholomorphic starlike mapping on the unit polydisk $\mathbb{P}^n$, then

$$w_i(z)|_{z_i=0} = 0, \quad i = 1, 2, \cdots, n.$$

In particular, $\varphi_i^{(2)}(z)|_{z_i=0} = 0, i = 1, 2, \cdots, n$, where $w_i(z), i = 1, 2, \cdots, n$ is defined by (5.3.4).

Proof. $w_i(z)$ is a holomorphic function of the variable z_i when $|z_i| < 1$ since $f(z)$ is biholomorphic in $\mathbb{P}^n$. By (5.3.4), $\frac{w_i(z)}{z_i}$ is bounded and holomorphic at a neighborhood of $z_i = 0$. By Riemann theorem, $\frac{w_i(z)}{z_i}$ is holomorphic at $z_i = 0$. By (5.3.3) and Herglotz representation theorem, we proved the Lemma.

Lemma 5.3.2. *Let $f(z)$ be a normalized biholomorphic starlike mapping on the unit polydisk $\mathbb{P}^n$, $w(z)$ be defined by (5.3.4), and let the Taylor expansion of $\frac{z_i}{w_i(z)}$ at $z = 0$ be*

$$(5.3.5) \qquad \frac{z_i}{w_i(z)} = 1 + \sum_{k=1}^{\infty} Q_i^{(k)}(z)$$

where $Q_i^{(k)}(z)$ is a homogeneous polynomal of the elements of z of degree k, then

$$(5.3.6) \qquad |Q_i^{(k)}(z)| \le 2\|z\|^k$$

holds for all $k = 2, 3, \cdots$, and $z \in P^n$.

Proof. Let $\varepsilon > 0$ be a positive small number. Theorem 5.3.2 tells us $\Re\{\varepsilon + \frac{w_i(z)}{z_i}\} > 0$ for all $z \in \mathbb{P}^n, 1 \le i \le n$, and $\Re\{\frac{(1+\varepsilon)z_i}{\varepsilon z_i + w_i(z)}\} > 0$. Hence, $g_{i,\varepsilon}(z) = \frac{(1+\varepsilon)z_i}{\varepsilon z_i + w_i(z)}$ is a holomorphic function on $\mathbb{P}^n$ with $g_{i,\varepsilon}(0) = 1$.

Let the Taylor expansion of $g_{i,\varepsilon}(z)$ at $z = 0$ be

$$g_{i,\varepsilon}(z) = 1 + \sum_{k=1}^{\infty} Q_{i,\varepsilon}^{(k)}(z)$$

where $Q_{i,\varepsilon}^{(k)}(z)$ is a homogeneous polynomial of the elements of z of degree k.

Let $\eta \in \mathbb{C}, |\eta| < 1$, then

$$\Re\left\{ g_{i,\varepsilon}\left(\frac{\eta z}{\|z\|} \right) \right\} \ge 0$$

and

$$g_{i,\varepsilon}\left(\frac{\eta z}{\|z\|} \right) = 1 + \sum_{k=1}^{\infty} \frac{\eta^k}{\|z\|^k} Q_{i,\varepsilon}^{(k)}(z).$$

Of course, $g_{i,\varepsilon}\left(\frac{\eta z}{\|z\|}\right)$ is a holomorphic function of η on the unit disk $\{\eta \in \mathbb{C}: |\eta| < 1\}$. By Lemma 1.2.4,

$$\left|\frac{Q_{i,\varepsilon}^{(k)}(z)}{\|z\|^k}\right| \le 2.$$

Thus,

$$|Q_{i,\varepsilon}^{(k)}(z)| \le 2\|z\|^k.$$

Let $\varepsilon \to 0$, then $Q_{i,\varepsilon}^{(k)}(z) \to Q_i^{(k)}(z)$. We have (5.3.6).

Proof of Theorem 5.3.1
At the neighborhood of $z = 0$, we have

$$z = \begin{pmatrix} \frac{z_1}{w_1(z)} & & 0 \\ & \ddots & \\ 0 & & \frac{z_n}{w_n(z)} \end{pmatrix} \begin{pmatrix} w_1(z) \\ \vdots \\ w_n(z) \end{pmatrix}.$$

Then

$$J_f(z)z = J_f(z) \begin{pmatrix} 1 + \sum_{k=1}^{\infty} Q_1^{(k)}(z) & & 0 \\ & \ddots & \\ 0 & & 1 + \sum_{k=1}^{\infty} Q_n^{(k)}(z) \end{pmatrix}$$
$$\times J_f^{-1}(z)f(z).$$

That is,

$$J_f(z)z - f(z) = J_f(z) \begin{pmatrix} \sum_{k=1}^{\infty} Q_1^{(k)}(z) & & 0 \\ & \ddots & \\ 0 & & \sum_{k=1}^{\infty} Q_n^{(k)}(z) \end{pmatrix}$$
(5.3.7) $$\times J_f^{-1}(z)f(z).$$

Using the property of Jacobian of f, we know that

$$J_f(z)z = z + \sum_{k=2}^{\infty} k\varphi^{(k)}(z)$$

and

$$J_f(z) = I + J_{\varphi^{(2)}}(z) + \cdots.$$

Thus, (5.3.7) implies

$$\sum_{k=2}^{\infty}(k-1)\varphi^{(k)}(z) = \left[(I + J_{\varphi^{(2)}}(z) + \cdots)\right.$$
$$\times \begin{pmatrix} \sum_{k=1}^{\infty} Q_1^{(k)}(z) & & 0 \\ & \ddots & \\ 0 & & \sum_{k=1}^{\infty} Q_n^{(k)}(z) \end{pmatrix}$$
$$\left. \times (I - J_{\varphi(2)}(z) + \cdots)\right] \cdot (z + \varphi^{(2)}(z) + \cdots)$$
$$= \left[\begin{pmatrix} Q_1^{(1)}(z) & & 0 \\ & \ddots & \\ 0 & & Q_n^{(1)}(z) \end{pmatrix} + \left\{ \begin{pmatrix} Q_1^{(2)}(z) & & 0 \\ & \ddots & \\ 0 & & Q_n^{(2)}(z) \end{pmatrix} \right.\right.$$
$$+ J_{\varphi(2)}(z) \begin{pmatrix} Q_1^{(1)}(z) & & 0 \\ & \ddots & \\ 0 & & Q_n^{(1)}(z) \end{pmatrix}$$
$$\left.\left. - \begin{pmatrix} Q_1^{(1)}(z) & & 0 \\ & \ddots & \\ 0 & & Q_n^{(1)}(z) \end{pmatrix} J_{\varphi(2)}(z) \right\} + \cdots \right]$$
$$\times (z + \varphi^{(2)} + \cdots).$$

Comparing the terms of some degree on both sides, we have

$$(5.3.8) \qquad \varphi^{(2)}(z) = \begin{pmatrix} Q_1^{(1)}(z) & & 0 \\ & \ddots & \\ 0 & & Q_n^{(1)}(z) \end{pmatrix} \begin{pmatrix} z_1 \\ \vdots \\ z_n \end{pmatrix}$$

and

$$2\varphi^{(3)}(z) = \begin{pmatrix} Q_1^{(1)}(z) & & 0 \\ & \ddots & \\ 0 & & Q_n^{(1)}(z) \end{pmatrix} \varphi^{(2)}(z)$$
$$+ \begin{pmatrix} Q_1^{(2)}(z) & & 0 \\ & \ddots & \\ 0 & & Q_n^{(2)}(z) \end{pmatrix} z$$

$$+ J_{\varphi^{(2)}}(z) \begin{pmatrix} Q_1^{(1)}(z) & & 0 \\ & \ddots & \\ 0 & & Q_n^{(1)}(z) \end{pmatrix} z$$

$$- \begin{pmatrix} Q_1^{(1)}(z) & & 0 \\ & \ddots & \\ 0 & & Q_n^{(1)}(z) \end{pmatrix} J_{\varphi^{(2)}}(z) z$$

$$= \begin{pmatrix} Q_1^{(2)}(z) & & 0 \\ & \ddots & \\ 0 & & Q_n^{(2)}(z) \end{pmatrix} z$$

(5.3.9)

$$+ J_{\varphi^{(2)}}(z)\varphi^{(2)}(z) - \begin{pmatrix} Q_1^{(1)}(z) & & 0 \\ & \ddots & \\ 0 & & Q_n^{(1)}(z) \end{pmatrix} \varphi^{(2)}(z)$$

by (5.3.8).

Using Lemma 5.3.2, (5.3.8) implies

(5.3.10) $$\|\varphi^{(2)}(z)\| \le 2\|z\|^2.$$

We may express $\varphi^{(2)}(z)$ as

(5.3.11) $$\varphi^{(2)}(z) = \begin{pmatrix} z_1(a_{11}z_1 + a_{12}z_2 + \cdots + a_{1n}z_n) \\ \vdots \\ z_n(a_{n1}z_1 + a_{n2}z_2 + \cdots + a_{nn}z_n) \end{pmatrix} = \begin{pmatrix} z_1 A_1 \\ \vdots \\ z_n A_n \end{pmatrix}$$

by Lemma 5.3.1, $\varphi_i^{(2)}(z)|_{z_i=0} = 0$, where $a_{ij}, i, j = 1, \cdots, n$ are constants and $A_i = a_{i1}z_1 + \cdots + a_{in}z_n, i = 1, 2, \cdots, n$.

From (5.3.8) and (5.3.11), we have

$$\begin{pmatrix} Q_1^{(1)}(z) & & 0 \\ & \ddots & \\ 0 & & Q_n^{(1)}(z) \end{pmatrix} = \begin{pmatrix} A_1 & & 0 \\ & \ddots & \\ 0 & & A_n \end{pmatrix}.$$

i.e., $Q_i^{(1)}(z) = A_i, i = 1, 2, \cdots, n$.

Using Lemma 5.3.2 again, the inequalities

$$|A_i| \le 2\|z\|, \quad i = 1, 2, \cdots, n, \quad z \in \mathbb{P}^n.$$

holds. It implies

$$(5.3.12) \qquad \sum_{j=1}^{n} |a_{ij}| \leq 2, \quad i = 1, 2, \cdots, n.$$

From (5.3.11), we have

$$J_{\varphi^{(2)}}(z) = \begin{pmatrix} 2a_{11}z_1 + a_{12}z_2 + \cdots + a_{1n}z_n, a_{12}z_1, \cdots, z_{1n}z_n \\ \vdots \\ a_{n1}z_n, a_{n2}z_n, \cdots, a_{n1}z_1 + a_{n2}z_2 + \cdots + 2a_{nn}z_n \end{pmatrix}$$

$$= \begin{pmatrix} A_1 & & 0 \\ & \ddots & \\ 0 & & A_n \end{pmatrix} + \begin{pmatrix} a_{11}z_1, a_{12}z_1, \cdots, a_{1n}z_1 \\ \vdots \\ a_{n1}z_n, a_{n2}z_n, \cdots, a_{nn}z_n \end{pmatrix}.$$

Thus,

$$J_{\varphi^{(2)}}(z)\varphi^{(2)}(z) - \begin{pmatrix} Q_1^{(1)}(z) & & 0 \\ & \ddots & \\ 0 & & Q_n^{(1)}(z) \end{pmatrix} \varphi^{(2)}(z)$$

$$= \left[\begin{pmatrix} A_1 & & 0 \\ & \ddots & \\ 0 & & A_n \end{pmatrix} + \begin{pmatrix} a_{11}z_1, & \cdots, & a_{1n}z_1 \\ \cdots & \cdots & \cdots \\ a_{n1}z_n, & \cdots, & a_{nn}z_n \end{pmatrix} \right.$$

$$\left. - \begin{pmatrix} A_1 & & 0 \\ & \ddots & \\ 0 & & A_n \end{pmatrix} \right] \varphi^{(2)}(z)$$

$$= \begin{pmatrix} a_{11}z_1\varphi_1^{(2)}(z) + \cdots + a_{1n}z_1\varphi_n^{(2)}(z) \\ \cdots \quad \cdots \quad \cdots \quad \cdots \\ a_{n1}z_n\varphi_1^{(2)}(z) + \cdots + a_{nn}z_n\varphi_n^{(n)}(z) \end{pmatrix}.$$

Using (5.3.10) and (5.3.12), we have

$$|a_{i1}z_i\varphi_1^{(2)} + \cdots + a_{in}z_i\varphi_n^{(2)}(z)| \leq \sum_{j=1}^{n} |a_{ij}| \cdot |z_i| \cdot |\varphi_j^{(2)}(z)|$$

$$\leq 2\|z\|^3 \sum_{j=1}^{n} |a_{ij}| \leq 4\|z\|^3.$$

By (5.3.6), we obtain

$$|Q_i^{(2)}(z)z_i| \leq 2\|z\|^3.$$

Using all these results to (5.3.9), we conclude

$$2\|\varphi^{(3)}(z)\| \leq 4\|z\|^3 + 2\|z\|^3 = 6\|z\|^3.$$

Thus,

$$\|\varphi^{(3)}(z)\| \leq 3\|z\|^3.$$

We have proved Theorem 5.3.1.

Of course, Theorem 5.3.1 is a generalization of Theorem 1.2.1 when $k = 2, 3$.

It is reasonable to make the following

Conjecture. *Let $f(z) = z + \sum_{k=2}^{\infty} \varphi^{(k)}(z)$ be a normalized starlike biholomorphic mapping on the unit polydisk $\mathbb{P}^n$, where the elements of $\varphi^{(k)}(z), k = 2, 3, \cdots$ are homogeneous polynomials of the elements of z of degree k, then*

$$\|\varphi^{(k)}(z)\| \leq k\|z\|^k$$

holds for all $k = 2, 3, 4, \cdots$, where $\| \ \| = \| \ \|_\infty$ is the norm of the unit polydisk $\mathbb{P}^n$.

That is, the Bieberbach conjecture is true when f is a normalized starlike biholomorphic mapping on the unit polydisk P^n.

In the case of the unit ball B^n in $\mathbb{C}^n$, we only have the following result. Let $f(z) = z + \sum_{k=2}^{\infty} \varphi^{(k)}(z)$ be a normalized starlike biholomorphic mapping on B^n where $f(z), z$ and $\varphi^{(k)}(z)$ are column vectors and each element of $\varphi^{(k)}(z)$ is a homogeneous polynomial of the elements of z of degree k, then $|\bar{z}'\varphi^{(2)}(z)| \leq 2$ holds. This result was obtained by Y. P. Huang [1].

Just like the convex mappings in several complex variables, there are a series of results about the starlike mappings in several complex variables also (cf., S. Gong [6]). For example, Theorem 5.3.2 mentioned the criterion for starlikeness for holomorphic mappings on the polydisk given by Suffridge. Actually, he proved more than this (T. J. Suffridge [1]).

Theorem 5.3.3. *Let B_p be defined by (5.2.1) and $f : B_p \to \mathbb{C}^n$ be a locally biholomorphic mapping with $f(0) = 0$, then f is a starlike mapping with respect to the origin if and only if $f(z) = wJ_f(z)'$, where $w \in \mathcal{P}_p$ and $\mathcal{P}_p$ is a family of holomorphic mappings for which $w \in \mathcal{P}_p$ is $w : B_p \to \mathbb{C}^n, w(0) = 0$ and*

$$(5.3.13) \qquad \Re e \sum_{j=1}^{n} \frac{w_j |z_j|^p}{z_j} \geq 0$$

holds for every $z \in B_p$.

In fact, we can be written these conditions as

$$(5.3.14) \qquad \Re e \sum_{j=1}^{n} \sum_{k=1}^{n} \frac{\partial \|z\|_p}{\partial z_j} \frac{\partial z_j}{\partial \xi_k} \xi_k \geq 0$$

where $\|z\|_p = (\sum_i |z_i|^p)^{\frac{1}{p}}$ is the norm of B_p, $f = (\xi_1, \cdots, \xi_n)$.

It is easy to check (5.3.14) is (5.3.3) when $p = \infty$, and (5.3.14) is (5.3.13) when $1 \leq p < \infty$.

We omit the detail of the proof of Theorem 5.3.3.

Of course, there are other interesting results about the criterion for star-likeness for holomorphic mappings in several complex variables.

Moreover, there are some other interesting results about the starlike mappings. For example, there are many interesting growth theorems for normalized biholomorphic starlike mappings in several complex variables. Here we state and prove the simplest one which was obtained by Barnard, FitzGerald and Gong (R. W. Barnard, C. H. FitzGerald and S. Gong [2]).

Theorem 5.3.4. *Let $f(z) : B^n \to \mathbb{C}^n$ be a normalized biholomorphic starlike mapping on the unit ball B^n in $\mathbb{C}^n$, then*

$$(5.3.15) \qquad \frac{|z|}{(1 + |z|)^2} \leq |f(z)| \leq \frac{|z|}{(1 - |z|)^2}.$$

holds for every $z \in B^n$. This estimate is precise.

The are many generalizations of this theorem.

We first prove the following Pfaltzgraff lemma (J. A. Pfaltzgraff [1]):

Lemma 5.3.3. *Let $w(z) : B^n \to \mathbb{C}^n$ be a normalized holomorphic mapping on B^n, satisfying the condition*

$$(5.3.16) \qquad \Re e \bar{z} w' \geq 0 \quad \text{when } z \in B^n,$$

then

$$(5.3.17) \qquad \frac{1 - |z|}{|z|(1 + |z|)} \cos \Xi \leq \frac{1}{|w(z)|} \leq \frac{1 + |z|}{|z|(1 - |z|)} \cos \Xi$$

where Ξ is the angle between z and $w(z)$.

Proof.   Fix $\zeta \in \partial B^n$, then by hypothesis,

$$\Re \sum_{j=1}^{n} \bar{\zeta}_j \bar{\xi} w_j(\zeta\xi) \geq 0$$

for $\xi \in D$. That is,

$$\Re \sum_{j=1}^{n} \bar{\zeta}_j \frac{w_j(\zeta\xi)}{\xi} \geq 0.$$

Let $P(\xi) = \sum_{j=1}^{n} \bar{\zeta}_j \frac{w_j(\zeta\xi)}{\xi}$, then $P(\xi)$ is a holomorphic function of ξ on D with a non-negative real part. By the normalization condition that $P(0) = 1$, we have

$$\frac{1 - |\xi|}{1 + |\xi|} \leq \Re P(\xi) \leq \frac{1 + |\xi|}{1 - |\xi|}$$

for any $\xi \in D$. We know that

$$\Re \sum_{j=1}^{n} \bar{\zeta}_j \frac{w_j(\zeta\xi)}{\xi} = \left\langle \frac{w(\zeta\xi)}{\xi}, \zeta \right\rangle = \frac{|w(\zeta\xi)|}{|\xi|} |\zeta| \cos\theta,$$

where θ is the angle between $\dfrac{w(\zeta, \xi)}{\xi}$ and ζ. Let $z = \zeta\xi$, then θ is Ξ. We have (5.3.17) by observing that $|\zeta| = 1$ and $|z| = |\xi|$.

Proof of Theorem 5.3.4.

Suppose $0 < r < 1$, and z_1 is a point on $|z| = r$, with $|f(z_1)| = \max_{|z|=r} |f(z)|$. The line segment which joins $f(z_1)$ and the origin lies in $f(|z| \leq r)$ because $f(z)$ is starlike. The inverse image of this line segment lies in the closure of B_r^n (a ball centered at the origin with radius r), and $|z|$ monotonically increases when z moves along the curve starting at the origin. The inverse image is an analytic curve, thus we may take arc length as the parameter of z, i.e., $z = z(s)$, then

(5.3.18) $$\frac{df(z(s))}{ds} = \sum_{j=1}^{n} \frac{\partial f}{\partial z_j} \frac{dz_j}{ds} \frac{dz}{ds}.$$

On the other hand, the image of $z(s)$ is a line segment, and the direction of this line segment is $f(z(s))$, thus we have a positive real-valued function $\lambda(z(s))$ such that

(5.3.19) $$\frac{df(z(s))}{ds} = \lambda(z(s)) f(z(s))$$

where f denotes a column vector $f = (f_1, \cdots, f_n)'$.

Since f is starlike, using Suffridge's criterion, Theorem 5.3.3, we have $w \in \mathcal{P}$, where w satisfies condition (5.3.16), and $f(z) = J_f w$. Substituting this into (5.3.19), we have

$$\frac{df(z(s))}{ds} = \lambda J_f w.$$

Comparing it to (5.3.18), we obtain

$$\frac{dz}{ds} = \lambda w.$$

We use arc length as the parameter of z, thus $\left|\dfrac{dz}{ds}\right| = 1$, i.e., $1 = \lambda|w|$. We may rewrite (5.3.19) as

(5.3.20) $$\frac{df(z(s))}{ds} = \frac{1}{|w(s)|} f(z(s)).$$

Let $g(s) = |f(z(s))|^2 = \overline{f(z(s))}' f(z(s))$, then

$$\frac{dg}{ds} = 2\Re\left\{\overline{f(z(s))}' \frac{df(z(s))}{ds}\right\}.$$

Substituting (5.3.20) into above equality, we have

$$\frac{dg}{ds} = 2\frac{1}{|w|}\Re\{\overline{f(z(s))}' f(z(s))\}\frac{2}{|w|}g(z(s)).$$

We obtain a differential equation

(5.3.21) $$\frac{dg}{g} = \frac{2}{|w|}ds.$$

Substituting the right inequality of Lemma 5.3.3 into the above equation, it becomes

$$\frac{d\log g}{ds} \leq 2\frac{1 + |z(s)|}{|z(s)|(1 - |z(s)|)} \cos\theta(s).$$

Since $\theta(s)$ is the angle between $z(s)$ and $\dfrac{dz(s)}{ds}$, we have $\dfrac{d|z(s)|}{ds} = \cos\theta(s)$ (if $z(s) \neq 0$). Integrating both sides of the preceding inequality with respect to s from s_0 to s_1 $(0 < s_0 < s_1 < 1)$, it follows that

$$\log g(s_1) - \log g(s_0) \leq \int_{s_0}^{s_1} \frac{2(1 + |z(s)|)}{|z(s)|(1 - |z(s)|)} \cos\theta(s)ds$$

$$= \int_{|z(s_0)|}^{|z(s_1)|} \frac{2(1+|z|)}{|z|(1-|z|)} d|z| = \int_{|z(s_0)|}^{|z(s_1)|} \left(\frac{2}{|z|} + \frac{4}{1-|z|} \right) d|z|$$
$$= 2\log|z(s_1)| - 4\log(1-|z(s_1)|) - [2\log|z(s_0)| - 4\log(1-|z(s_0)|)].$$

Let $s_0 = \varepsilon$, then $|z(s_0)| = \varepsilon + o(\varepsilon)$, $g(s_0) = \varepsilon^2 + o(\varepsilon^2)$, since f is normalized. This implies that

$$g(s_1) \leq \left\{ \frac{|z(s_1)|^2}{(1-|z(s_1)|)^4} \right\} \frac{\varepsilon^2 + o(\varepsilon^2)}{\varepsilon^2 + o(\varepsilon^2)}.$$

Choose s_1 such that $z(s_1) = z$, and let $\varepsilon \to 0$, then we obtain the right inequality of (5.3.16).

We obtain the left inequality if we apply the left inequality of Lemma 5.3.3 to (5.3.21).

This estimate of (5.3.15) is precise. For example, when $n = 2$, if we let $f(z) = \left(\frac{z_1}{(1-z_1)^2}, \frac{z_2}{(1-z_2)^2} \right)'$ and $z = (z_1, 0)$, then the equalities hold on both sides of (5.3.15). Actually, if we let $f = \left(\frac{z_1}{(1-z_1)^2}, f_2(z_2) \right)'$, where $f_2(z_2)$ is a normalized starlike mapping on D, then there exist points such that the equalities hold on both sides of (5.3.15). It is not difficult to verify that f is a normalized starlike mapping.

We have proved Theorem 5.3.4.

From Theorem 5.3.4, we obtain the covering theorem:

Corollary 5.3.1.   *With the same assumptions as in Theorem 5.3.4, $f(B^n)$ contains a ball centered at the origin with radius $\frac{1}{4}$. The value $\frac{1}{4}$ is precise.*

The preceding examples attain the value $\frac{1}{4}$.

Let $f : B^n \to \mathbb{C}^n$, be a holomorphic mapping on B^n in $\mathbb{C}^n$, then f is k-fold symmetric if $\exp\left(\frac{-2\pi i}{k} \right) f(e^{\frac{2\pi i}{k}} z) = f(z)$ for all $z \in B^n$. In the process of the proof of Theorem 5.3.4, we know that the $w \in \mathcal{P}$ in Suffridge's criterion is k-fold symmetric if f is k-fold symmetric, and $P(\xi)$ in Lemma 5.3.3 is k-fold symmetric. We have

$$\frac{1-|\xi|^{\frac{1}{k}}}{1+|\xi|^{\frac{1}{k}}} \leq \Re P(\xi) \leq \frac{1+|\xi|^{\frac{1}{k}}}{1-|\xi|^{\frac{1}{k}}}.$$

Using this estimate in (5.3.21), it follows that:

Corollary 5.3.2. *Let $f(z) : B^n \to \mathbb{C}^n$ be a normalized biholomorphic starlike k-fold symmetric mapping, then the inequalities*

$$(5.3.22) \qquad \frac{|z|}{(1 + |z|^k)^{\frac{2}{k}}} \leq |f(z)| \leq \frac{|z|}{(1 - |z|^k)^{\frac{2}{k}}}$$

hold for every $z = (z_1, \cdots, z_n) \in B^n$. The estimates are precise.

Obviously, the mapping $\left(\dfrac{z_1}{(1 - z_1^k)^{2/k}}, \dfrac{z_2}{(1 - z_2^k)^{2/k}} \right)$ makes the equalities of (5.3.22) hold at some points. This is a normalized biholomorphic starlike k-fold symmetric mapping.

As we know that if $\Omega \subset \mathbb{C}^n$ is a domain, Ω is called circular if $w \in \Omega$ implies $\xi w \in \Omega$ for any ξ such that $|\xi| = 1$. Ω is called a balanced domain if Ω is circular, $0 \in \Omega$, and Ω is starlike with respect to the origin.

Corollary 5.3.3. *Let $f : B^n \to \mathbb{C}^n$ be a normalized biholomorphic mapping on B^n, and let $f(B^n)$ be a balanced domain, then $f(B^n) = B^n$.*

Because $f(B^n)$ is a balanced domain, (5.3.22) holds for any k. Let $k \to \infty$, then we have $|f(z)| = |z|$. By the normalization condition, it follows that $f(z) = z$, i.e., $f(B^n) = B^n$.

As a consequence of Corollary 5.3.3, we obtain the well-known Poincaré Theorem:

Corollary 5.3.4. *The unit ball B^n in $\mathbb{C}^n$ and the polydisk in $\mathbb{C}^n$ are not biholomorphically equivalent when $n > 2$.*

Applying an affine transformation, we may assume that the polydisk in $\mathbb{C}^n$ is $\mathbb{P}^n = \{w = (w_1, \cdots, w_n) : |w_k| \leq r_k, k = 1, \cdots, n\}$. Assume there exists a biholomorphic mapping which maps B^n onto $\mathbb{P}^n$. After a Möbius transformation, we may assume that $f = 0$ at the origin, and we may diagonalize the Jacobian of f at the origin. After we change the scale of the coordinates, the Jacobian of f at the origin is the identity matrix, I . Then f is a normalized biholomorphic mapping and $f(B^n) = \mathbb{P}^n$. This contradicts Corollary 5.3.3, which means that the mapping does not exist.

REFERENCES

Aharonov, D.

[1] Proof of the Bieberbach conjecture for a certain class of univalent functions, *Isreal J. Math.*, 8(1970), 103-104.

[2] On the Bieberbach conjecture for functions with small second coefficient, *Isreal J. Math.*, 15(1973), 137-139.

[3] Bazilevich theorem and the growth of univalent functions, *Complex Analysis II*, Lecture Notes in Math. 1275, (Ed. C. A. Berenstein) pp. 1-9.

Ahlfors, L. V.

[1] *Complex Analysis*, 3rd edition, 1979, McGraw-Hill Book Co.

[2] *Conformal invariants, Topics in geometric function theory*, 1973, McGraw-Hill Book Company.

Aleksandrov, I. A. and Milin, I.M.

[1] The Bieberbach conjecture and logarithmic coefficients of univalent function, *Izv. Vyssh. Uchebn. Zaved. Mat.* 1989, pp. 3-15, (in Russian).

Andersén, E.

[1] Volume-preserving automorphisms of C^n, *Complex Variables*, 14(1990) 223-235.

Andersén, E. and Lempert, L.

[1] On the group of holomorphic automorphisms of C^n, *Invent. Math.*, 110(1992), 371-388.

Askey, R. and Gasper, G.

[1] Positive Jacobi polynomial sum II, *Amer. J. math.*, 98(1976), 709-737.

[2] Inequalities for polynomials, *The Bieberbach conjecture* (West Lafayette, Ind.) (1985), 7-32, Amer. Math. Soc. Providence RI 1986.

183

Baernstein, A.

[1] Integral means, univalent functions and circular symmetrization, *Acta Math.*, 133(1974), 139-169.

Baranova, V. A.

[1] An estimate of the coefficient C_4 of univalent function depending on $|C_2|$, *Math. Notes*, 12(1972), 510-512.

Barnard, R. W., FitzGerald, C. H. and Gong. S.,

[1] Distortion theorem for biholomorphic mapping in $\mathbb{C}^2$, *Tran. Amer. Math. Soc.*, 344(1994), 907-924.

[2] The growth and $\frac{1}{4}$-theorems for starlike mappings in $\mathbb{C}^n$, *Pacific Jour. of Math.* 150(1991), 13-22.

Bazilevich, I. E.

[1] Coefficient dispersion of univalent functions, *Mat. Sb.*, 68(110) (1965), 549-560.(in Russian)

[2] On a univalence criterion for regular functions and the dispersion of their coefficients, *Mat. Sb.*,74(116) (1967), 133-146. (in Russian)

[3] On a case of integrability by quadratures of the equation of Löwner-Kufarev, *Mat. Sb.*, 37(79)(1955), 471-476.(in Russian)

[4] Improvement of the estimates of coefficients of Univalent functions, *Mat. Sb.*, 22(64), (1948) 381-390. (in Russian)

[5] On distortion theorems in the theory of univalent functions, *Mat. Sb.*, 28(70)(1951), 283-292.(in Russian)

Bernadi, S. D.

[1] *Bibliography of Schlicht Functions*, Mariner Publishing Company, Inc, 1982.

Bieberbach, L.

[1] Über die Koeffizienten derjenigen Potenzreihen, welche eine schlichte Abbildung des Einheitskreises vermitteln, *S. B. Preuss Akad. Wiss.*, (1916), 940-955.

Biernacki, M.

[1] Sur les coefficients Tayloriens des fonctions univalents, *Bull. Acad. Polon. Sci.*, 4(1956), 5-8.

Bishouty, D. H.

[1] The Bieberbach conjecture for univalent functions with small second coefficients, *Math. Z.*, 149(1976), 183-187.

[2] The Bieberbach conjecture for restricted initial coefficients, *Math. Z.*, 182(1983), 149-158.

de Branges, L.

[1] A proof of the Bieberbach conjecture, *Acta Math.*, 154(1985), 137-152.
[2] A proof of the Bieberbach conjecture, preprint E-5-84, Leningrad Branch of the V. A. Steklov Mathematical Institute, 1984.
[3] *Square summable power series*, 2nd edition, to appear.
[4] Powers of Riemann mapping functions, *The Bieberbach conjecture* (West Lafayette, Ind.)(1985) 51-67, Amer. Math. Soc., Proridence, RI 1986.
[5] Underlying concepts in the proof of the Bieberbach conjecture, *Proceedings of the International Congress of Mathematicians*, 1986, pp. 25-42, Berkelay, California, 1986.
[6] Das mathematische Erbe von Ludwig Bieberbach (1886-1982), *Nieuw Arch Wisk*, (4) 9 (1991), 366-370.
[7] Unitary linear systems whose transfer functions are Riemann mapping functions, *Integral Equations and Operation Theory*, 19(1986), 105-124.

Cartan, H.

[1] Sur la possibilité d'étendre aux fonctions de plusieurs variables complexes la theorie des fonctions univalents, *Lecons sur les Fonctions Univalents ou Multivalents*, by P. Montel, Gauthier-Villar, 1993, pp. 129-155.

Charzynski, Z. and Schiffer, M.

[1] A new proof of the Bieberbach conjecture for the fourth coefficient, *Arch Rational Mech. Anal.*, 5(1960), 187-193.

Chen, K. K.

[1] On the theory of schlicht functions, *Proc. Imp. Acad. Japan*, 9(1933), 465-467.

Clausen, Th.

[1] Beitrag zur Theorie der Reihen, *J. für die reine und angewandte Math.* 3 (1828), 92-95.

Conway, J. B.

[1] *Functions of one complex variable, II*, Graduate texts in Math. 159, Springer-Verlag, 1995.

Dieudonné, J.

[1] Sur les fonctions univalents, *C. R. Acad Sci.* Paris, 192(1931), 1148-1950.

Duren, P. L.
[1] *Univalent Functions*, Springer-Verlag, 1983.

Ehrig, G.
[1] The Bieberbach conjecture for univalent functions with restricted second coefficients, *J. London Math. Soc.*, 8(1974), 355-360.
[2] Coefficient estimates concerning the Bieberbach conjecture, *Math. Z.*, 140(1974), 111-126.

Ekhad, S. B. and Zeilbergen, D.
[1] A high-school algebra, "formal calculus", proof of Bieberbach conjecture [after L. Weinstein], Jerusalem combinatorics'93, pp. 113-115, *Contemp. Math.*, 178, Amer. Math. Soc., Providence, RI, 1994.

Fekete, M. and Szegö, G.
[1] Eine Bemerkung über ungerade schlichte Funktionen, *J. London Math. Soc.*, 8 (1933), 85-89.

FitzGerald, C. H.
[1] Quadratic inequalities and coefficient estimates for schlicht functions, *Arch. Rational Mech. Anal.* 46(1972), 356-368.
[2] Quadratic inequalities and analytic continuation, *Journal D'analyse Math ématique*, 31(1977), 19-47.
[3] Geometric function theory in one and several complex veriables: parallels and problems, *Complex analysis and its applications* (C. C. Yang, G. C. Wen, K. Y. Li and Y. M. Chiang, Ed) Pitman Research Notes in Math. Series 305, (1994) pp.14-25, Longman Scientific and Technical.
[4] Coefficient bounds for holomorphic convex mappings in complex variables, preprints, 1995.
[5] The Biebarbach conjecture; retrospective, *Notices AMS*, 32(1985) pp.2-6.

FitzGerald, C. H. and Horn, R. A.
[1] On the structure of Hermitian-symmetric inequalities, *J. London Math. Soc.*(2), 15(1977), 419-430.

FitzGerald, C. H. and Pommerenke, Ch.
[1] The de Branges theorem on univalent functions, *Tran. Amer. Math. Soc.*, 290 (1985), 683-690.

Garabedian, P. R. and Schiffer, M.
[1] A proof of the Bieberbach conjecture for the fourth coefficients, *J. Rational Mech Anal.*, 4(1955), 427-465.

Gasper, G.

[1] A short proof of an inequality used by de Branges in his proof of Bieber-bach, Robertson and Milin Conjectures, *Complex Variables. Theory Appl.* 7 (1986), 45-50.

Goluzin, G. M.

[1] *Geometric Theory of Functions of a Complex Variable*, 2nd ed., Izdat. "Nauka": Moscow, 1966; English transl. Amer. Math. Soc., 1969.

[2] On distortion theorems in the theory of conformal mappings, *Mat. Sb.*, 1 (43)(1936), 127-135. (in Russian)

[3] A method of variation in conformal mapping II, *Mat. Sb.*, 21(63) (1947), 83-117. (in Russian)

[4] On distortion theorems and coefficients of univalent functions, *Mat. Sb.*, 23(65)(1948), 353-360. (in Russian).

[5] On the theory of univalent functions, *Mat. Sb.*, 29 (71)(1951), 197-208. (in Russian)

[6] On the coefficients of univalent functions, *Mat. Sb.*, 22 (64)(1948), 373-380.(in Russian)

[7] On distortion theorems and coefficients of univalent functions, *Mat. Sb.*, 19(61)(1946), 183-202. (in Russian)

Gong, Sheng

[1] Contributions to the theory of schlicht functions I, Distortion theorem, *Scientia Sinica*, 4(1955), 229-249; II, The coefficient problem, *Scientia Sinica*, 4(1955),359-373.

[2] A simple proof of Bieberbach conjecture for the sixth coefficients, *Scientia Sinica, Mathematics* (1979), 1157-1170. (in Chinese)

[3] Coefficient inequalities and geometric inequalities, *Chinese Science Bulletin*, 31(1986), 1209-1212.

[4] The $\kappa(t)$ function in Löwner differential equations, *Acta Mathematica Sinica*, 3(1953) 225-230.(in Chinese)

[5] The Bieberbach conjecture for univalent functions with restricted second coefficients, *Scientia Sinica*, Mathematics(I)(1979), 202-214.

[6] *Convex and starlike mappings in several complex variables*, Kluwer Academic Publishers, 1998.

Gong, Sheng and Liu, Taishun

[1] The growth theorem of biholomorphic convex mappings on B_p, *Chinese Querterly Journal of Math.* 6(1991), 78-82.(in Chinese)

Gong. S., Wang, S. K. and Yu, Q.H

[1] Biholomorphic convex mappings of ball in $\mathbb{C}^n$, *Pacific Jour. of Math.*, 161(1993) 287-306.

Gong, Sheng and Yan, Zhimin

[1] A remark on Möbius transformations III, *Chinese Quarterly Journal of Mathematics*, 1(1986), 33-40. (in Chinese)

Gong, Sheng and Zheng Xuean

[1] Distortion theorem for biholomorphic mappings in transitive domains I. *International symposium in memory of L. K. Hua, Vol. II*, Springer-Verlag, 1991, 111-122.

Goodman, A. W.

[1] *Univalent functions*, Vol I, II, Mariner Publishing Co., Tampa Florida, 1983.

Grinspan, A. Z.

[1] Improved bounds for the difference of the moduli of adjoint coefficients of univalent functions, in "*Some Questions in the modern Theory of Functions*" (Sib. Int. Mat; Novosibirsk, 1976), 41-45. (in Russian)

Grunsky, H.

[1] Koeffizientenbedingungen för schlicht abbildende meromorphe Funktionen, *Math. Z.*, 45(1939), 29-61.

Hamilton, D. H.

[1] On Littlewood's conjecture for univalent functions, *Proc. Amer. Soc.*, 86(1982), 32-36.

Hayman, W. K.

[1] *Multivalent Functions*, Cambridge University Press, 2nd edition, 1994.
[2] The asymptotic behaviour of p-valent functions, *Proc. London Math. Soc.*, 5(1955), 257-284.
[3] On successive coefficient estimates of univalent functions, *J. London Math. Soc.*, 38(1963), 228-243.

Helton, J. W. and Weening, F.

[1] Some systems theorems arising from the Bieberbach conjecture, *J. Nonlinear and Robust Control*, to appear.

Henrici, P.

[1] *Applied and Computational Complex Analysis Vol III*, Wiley, New York, 1986.

Horowitz, D.

[1] A refinement for coefficient estimates of univalent functions, *Proc. Amer. Math. Soc.*, 54(1976), 176-178.

[2] A further refinement for coefficient estimates of univalent functions, *Proc. Amer. Math. Soc.*, 71 (1978), 217-221.

Hu, Ke

[1] Coefficients of odd univalent functions, *Proc. of Amer. Math. Soc.*, 96(1986), 183-186.

[2] Adjacent coefficients of univalent functions, *Chinese Annals of Math.* 10B (1989), 38-42.

Hua, Loo-Keng

[1] *Harmonic Analysis of Functions of Several Complex Variables in the Classical Domains*, Transl. of Math. Monographs Vol.6, Amer. Math. Soc. 1963.

Huang, Y. P.

[1] Personal communication.

Ilina, I. P.

[1] On the relative growth of adjoint coefficients of univalent functions, *Math. Notes*, 4(1968), 918-922.

[2] Estimates for the coefficients of univalent functions in terms of the second coefficient, *Math. Notes*, 13(1973), 215-218.

Jahangiri, M.

[1] Personal communication.

Jenkins, J. A.

[1] *Univalent Functions and Conformal Mapping*, Springer-Verlag, 1958.

Kazarinoff, N.

[1] Special functions and the Bieberbach conjecture, *Amer. Math. Society Monthly*, 95(1988), 689-696.

Keogh F. R. and Merkes E. P.

[1] A coefficient inequality for certain class of analytic functions, *Proc. Amer. Math. Soc.* 20(1969), 8-12.

Kikuchi, K

[1] Starlike and convex mappings in several complex variables, *Pacific Jour. of Math.*, 44(1973), 569-580.

Koapf. W.

[1] Von der Bieberbachschen Vermutung zum Satz von de Branges sowie der Beweisvariante von Weinstein, *Jahrbuch Überblicke Mathematik*, 1994, 175-193.

Koornwinder, T. H.

[1] A group theoretic interpretation of the last part of de Brangers' proof of the Bieberbach conjecture, *Complex Variables Theory, Appl.* 6(1986), 309-321.

Korevaar, J.

[1] Ludwig Bieberbach's conjecture and its proof by Louis de Branges , *Amer. Math. Monthely,* 93(1986), 504-514.

Kufarev, P. P.

[1] On one-parameter families of analytic functions, *Mat. Sb.*, 13(55)(1943), 87-118. (in Russian)

[2] A theorem on solutions of a differential equation, *Uchen. Zap. Tomsk. Gos. Univ.*, 5 (1947), 20-21 (in Russian)

Landau, E.

[1] Über schlichte Funktionen, *Math, Z.*, 30(1929), 635-638.

[2] Einige Bemerkungen über schlichte Abbildung, *Jber. Deutsh. Math. Verein.*, 34(1925-26), 239-243.

Lebedev, N. A.

[1] *The Area Princiole in the Theory of Univalent Functions*, Izdat. "Nauka": Moscow, 1975. (in Russian)

Lebedev, N. A. and Milin, I. M.

[1] On the coefficients of certain classes of analytic functions, *Mat. Sb.*, 28(70)(1951),359-400. (in Russian)

[2] An inequality, *Vestnik Leningrad. Univ.*, 20(1965), no.19, 157-158. (in Russian)

Leeman, G. B.

[1] The seventh coefficient of odd symmtric univalent functions, *Duke Math. J.*, 43(1976), 301-307.

Leung, Y. J.
[1] Successive coefficients of starlike functions, *Bull. London Math. Soc.*, 10(1978), 193-196.

Levin, V. I.
[1] Some remarks on the coefficients of schlicht functions, *Proc. London Math. Soc.*, 39(1935), 467-480.

Littlewood, J. E.
[1] On inequalities in the theory of functions, *Proc. London Math. Soc.*, 23(1925),481-519.

Littewood, J. E. and Paley, R. E. A. C.
[1] A proof that an odd schlicht function has bounded coefficients,*J. London Math. Soc.*, 7(1932), 167-169.

Liu, T. S.
[1] Personal communication.
[2] The distortion theorem for biholomorphic mappings in $\mathbb{C}^n$, Preprient, 1989.
[3] On the estimate of the coefficients of starlike mappings on polydisc, Preprints, 1997.
[4] The growth theorems and covering theorems for biholomorphic mappings on classical domains, University of Science and Technology of China, Doctoral Dissertation, 1989.

Liu, Taishun and Ren, Guangbin.
[1] The growth theorem of convex mappings on bounded convex circular domains, Science in China, Series A.41 (1998) 123-130.

Löwner, K. (C. Loewner)
[1] Untersuchungen über schlichte konforme Abbildungen des Einheitskreises, I. *Math. Ann.* 89(1923), 103-121.
[2] Untersuchungen über die Verzerrung bei konformen Abbildungen des Einheitskreises $|z| < 1$, die durch Funktionen mit nicht verschwindender Ableitung geliefertwerden, *Ber. Verh. Sächs. Ges. Wiss. Leipzig*, 69(1917), 89-106.

Milin, I. M.
[1] *Univalent Functions and Orthonormal Systems*, English transl. Amer. Math. Soc., 1977.
[2] Estimation of coefficients of univalent functions, *Soviet Math. Dokl.* 6 (1965),196-198.

[3] On the coefficients of univalent function, *Soviet Math. Dokl.*, 8(1967), 1255-1258.

[4] Adjoint coefficients of univalent functions, *Soviet Math. Dokl.*, 9(1968), 762-765.

[5] Hayman's regularity theorem for the coefficients of univalent functions, *Soviet Math. Dokl.* 11(1970), 724-728.

Milin V. I.

[1] Estimate of the coefficients of odd univalent function, *Metric question of the theory of functions* (G. D. Surorov, ed.) "Naukova Dumka" Kiev, 1980, pp. 78-86 (in Russian).

Nahari, Z.

[1] On the coefficients of univalent functions, *Proc. Amer. Math. Soc.*, 8(1957),291-293.

[2] A proof of $|a_4| \leq 4$ by Löwner's method, *Proceedings of the Symposium on Complex Analysis*, Canterburg, 1973, London Math. Soc. Lecture Notes Series No. 12., Cambridge University Press, 1974, 107-110.

Nevanlinna, R.

[1] Über die konforme Abbildung von Sterngebieten, *Översikt av Finska Vetenskaps Soc. Förh.*, 63(A), no. 6 (1920-21), 1-21.

Nikol'skii, N. K and Vasyunin, V. I.

[1] Quasiorthogonal decompositions with respect to complementary metrics, and estimates of univalent functions, *Leningrad Math. J.* 2(1991)691-764.

[2] Operator-Valued meansures and coefficients of univalent functions, *St. Petersburg Math. J.* 3(1992) 1199-1270.

Ozawa, M.

[1] On the Bieberbach conjecture for the sixth coefficients, *Arch. Rational Mach. Anal.*, 21(1969), 97-128.

Pederson, R. N.

[1] A proof of the Bieberbach conjecture for the sixth coefficients, *Arch. Rational Mech. Anal.*, 31(1968), 331-351.

Peterson, R. N. and Schiffer, M.

[1] A proof of the Bieberbach conjecture for the fifth coefficient, *Arch. Rational Mech. Anal.*, 45(1972), 161-193.

Pfaltzgraff, J. A.
[1] Subordination chains and univalence of holomorphic mappings in $\mathbb{C}^n$, *Math. Annalen*, 210(1974),55-68.

Pfaltzgraff, J. A. and Suffridge T. J.
[1] Linear invariant, order and convex maps in $\mathbb{C}^n$, Research report, 96-6, 1996, Univ. of Kentucky.

Pommerenke, Ch.
[1] *Univalent Functions*, Vanderhoeck and Puprecht; Göttingen, 1975.
[2] Linear-invariante Familien analytischer Functioner I, *Math. Ann.*, 155 (1964), 108-154; II, *Math. Ann.* 156 (1964), 226-264.
[3] The Bieberbach conjecture, *Math. Intelligencer*, 7(1985) 23-25, 32.
[4] Probleme aus der Funktionen-theorie, *Jber. Deutsch. Math.-Verein*,73 (1971) 1-5.

Reade, M. O.
[1] On close-to-convex univalent functions, *Michigan Math. J.*, 3(1955-56), 59-62.

Robertson, M. S.
[1] The generalized Bieberbach conjecture for subordinate functions, *Michigan Math. J.*, 12(1965), 421-429.

Rogosinski, W.
[1] Über positive harmonische Entwicklungen und typisch-reelle Potenzreihen, *Math. Z.*, 35(1932), 93-121.
[2] On subordinate functions, *Proc. Cambridge Philos. Soc.*, 35(1939), 1-26.
[3] On the coefficients of subordinate functions, *Proc. London Math. Soc.* 48(1943), 48-82.

Rosenblum, M. and Rovnyak, J.
[1] *Topics in Hardy classes and univalent functions*, Birkhäuser Verlag, 1994.

Rosay, J. P. and Rudin, W.
[1] Holomorphic maps from $\mathbb{C}^n$ to $\mathbb{C}^n$, *Tran. Amer. Math. Soc.*, 310(1988), 467-486.

Rovnyak, J.
[1] Coefficient estimates for Riemann mapping functions, *J. d'Analyse Mathématique* 52(1989) 53-93.

Schaeffer, A. O. and Spencer, D. C.

[1] *Coefficient Regions for Schlicht Functions*, Amer. Math. Soc. Colloq. Publ., vol. 35, 1950.

[2] The coefficients of schlicht functions, *Duke Math. J.*, 10(1943), 611-635.

Schober, G.

[1] *Univalent Functions-Select Topics*, Lecture Notes in Math. No.478, Springer-Verlag, 1975.

Sheil-Small, T.

[1] On the convolution of analytic functions, *J. Reine Angew. Math.*, 258 (1973), 137-152.

Suffridge, T. J.

[1] The principle of subordination applied to function of several variables, *Pacific Jour. of Math.* 33(1970), 241-248.

[2] Biholomorphic mappings of ball onto convex domains, *Abstracts of papers presented to AMS* 11(66)(1990), p.46.

Sza'sz, O.

[1] Über Funktionen, die den Einheitskreise schlicht abbilden. *Jber. Deutsch. Math.-Verein.*, 42(1933), 73-75.

Thomas, C. R.

[1] Extensions of classical results in one complex variable to several complex variables, university of California, San Diego, Doctoral Dissertation, 1991.

Todorov, P. G.

[1] A simple proof of the Bieberbach conjecture, Serdica 19(1993) no. 2-3, 204-214; Acad. Roy. Belg. Bull. Cl. Sci (6) (1992) no. 12, 335-346.

Weinstein, L.

[1] The Bieberbach Conjecture, *International Math. Research Notices*, 5 (1991); *Duke Math. J.* 64(1991), 61-64.

Whittaker, E. T. and Watson, G. N.

[1] *A course of Modern Analysis*, 4th edition, 1962, Cambridge.

Wilf, H. S.

[1] A footnote on two proofs of the Bieberbach-De Branges theorem, *Bull. London Math. Soc.*, 26(1994), 61-63.

Ye, Z. Q.

[1] In successive coefficients of univalent functions, *Jour. of Jiangxi Normal Univ. (Science edition)*, 1(1985) pp. 24-33.

LIST OF SYMBOLS

INDEX